LIFE SCIENCES RESEARCH AND DEVELOPMENT

STEROLS

TYPES, CLASSIFICATION AND STRUCTURE

LIFE SCIENCES RESEARCH AND DEVELOPMENT

Additional books and e-books in this series can be found on Nova's website under the Series tab.

LIFE SCIENCES RESEARCH AND DEVELOPMENT

STEROLS

TYPES, CLASSIFICATION AND STRUCTURE

SCOTT JIMENEZ
EDITOR

Copyright © 2020 by Nova Science Publishers, Inc.

All rights reserved. No part of this book may be reproduced, stored in a retrieval system or transmitted in any form or by any means: electronic, electrostatic, magnetic, tape, mechanical photocopying, recording or otherwise without the written permission of the Publisher.

We have partnered with Copyright Clearance Center to make it easy for you to obtain permissions to reuse content from this publication. Simply navigate to this publication's page on Nova's website and locate the "Get Permission" button below the title description. This button is linked directly to the title's permission page on copyright.com. Alternatively, you can visit copyright.com and search by title, ISBN, or ISSN.

For further questions about using the service on copyright.com, please contact:
Copyright Clearance Center
Phone: +1-(978) 750-8400   Fax: +1-(978) 750-4470   E-mail: info@copyright.com.

NOTICE TO THE READER

The Publisher has taken reasonable care in the preparation of this book, but makes no expressed or implied warranty of any kind and assumes no responsibility for any errors or omissions. No liability is assumed for incidental or consequential damages in connection with or arising out of information contained in this book. The Publisher shall not be liable for any special, consequential, or exemplary damages resulting, in whole or in part, from the readers' use of, or reliance upon, this material. Any parts of this book based on government reports are so indicated and copyright is claimed for those parts to the extent applicable to compilations of such works.

Independent verification should be sought for any data, advice or recommendations contained in this book. In addition, no responsibility is assumed by the Publisher for any injury and/or damage to persons or property arising from any methods, products, instructions, ideas or otherwise contained in this publication.

This publication is designed to provide accurate and authoritative information with regard to the subject matter covered herein. It is sold with the clear understanding that the Publisher is not engaged in rendering legal or any other professional services. If legal or any other expert assistance is required, the services of a competent person should be sought. FROM A DECLARATION OF PARTICIPANTS JOINTLY ADOPTED BY A COMMITTEE OF THE AMERICAN BAR ASSOCIATION AND A COMMITTEE OF PUBLISHERS.

Additional color graphics may be available in the e-book version of this book.

Library of Congress Cataloging-in-Publication Data

Names: Jimenez, Scott, editor.
Title: Sterols: Types, Classification and Structure
Description: New York: Nova Science Publishers, [2019] | Series: Life Sciences Research and Development | Includes bibliographical references and index.
Identifiers: LCCN 2019957629 (print) | ISBN 9781536172317 (paperback) | ISBN 9781536172324 (adobe pdf)

Published by Nova Science Publishers, Inc. † New York

CONTENTS

Preface		vii
Chapter 1	Identification and Quantification of Sterols in Coral Reef Food Webs *Jorge A. Del Angel-Rodríguez, Laura Carreón-Palau and Christopher C. Parrish*	1
Chapter 2	Freshwater Sponge Sterols *Iuri Bezerra de Barros, Glaucia Cristina Manço da Costa Bolson and Valdir Florencio da Veiga Junior*	43
Chapter 3	Cultivar Effect on Sterol Composition of Virgin Olive Oil *Bechir Baccouri, Hedia Manai-Djebali and Leila Abaza*	59
Chapter 4	Investigation of Sterol Compounds of Virgin Olive Oil from New Cultivars Obtained through Uncontrolled Crossings *Bechir Baccouri and Imene Rajhi*	73

Chapter 5	Sterols of Virgin Argan Oil: Comparison with Olive Oil *Bechir Baccouri and Imene Rajhi*	85
Chapter 6	Use of Phytosterols as a Tool for the Authenticity Assessment of Virgin Olive Oil: Protection of the Olive Oil Market *Imen Oueslati, Hedia Manai-Djebali and Ridha Mhamdi*	97
Chapter 7	Analysis of Phytosterols: A Survey of the Analytical Protocols in Use *Svetlana M. Momchilova and Boryana M. Nikolova-Damyanova*	117
Index		137
Related Nova Publications		145

PREFACE

Sterols: Types, Classification and Structure describes the methodology used to determine sterol content in coral reef food webs, and how sterol data is used in trophic ecology studies. The authors briefly explain the basics of gas chromatography and mass spectrometry as necessary tools to confidently separate, identify, and quantify sterols.

Sterols from different sponge families have been described, however, few studies have been conducted involving the sterols of the freshwater sponge compared to those of the marine environment. The number of sterols present varies with the species, and this characteristic make it possible to use this data as a chemosystematic marker.

In olive oil, sterols constitute the majority of the unsaponifiable fraction. In recent years, there has been increased interest in the sterols of olive oil due to their health benefits and their importance to virgin olive oil quality. Thus, the authors discuss recent findings concerning the effect of cultivars on the sterolic profile of extra virgin olive oil.

The subsequent study aims to contribute to the optimisation and valorisation of virgin olive oil quality in the world olive-producing areas. This work was carried out on the study of virgin olive oil from two new olive varieties obtained through uncontrolled crossings.

The sterolic fraction of argan oil is also compared to that of olive oil. The total phytosterol content ranged from 1700.80 mg/kg in chemlali oil to

150.40mg/kg in argan oil. In contrast to chemlali oil in which β-sitosterol is predominant, the major sterols detected in argan oil is schottenol and spinasterol.

In closing, the authors examine the analysis of phytosterols, a multistage procedure that includes extraction, isolation/purification as a group of related compounds and a chromatographic technique for separation, identification, and quantification.

Chapter 1 - This chapter describes in detail the methodology used to determine sterols in coral reef food webs and how the authors use the sterol data in trophic ecology studies. The authors briefly explain the basics of gas chromatography and mass spectrometry as necessary tools to confidently separate, identify, and quantify sterols. The authors describe the steps and reactions required to form sterol-trimethylsilyl ethers. This is needed to improve the chromatographic separation and identification via mass spectrometry. A main goal of this chapter is to facilitate the interpretation of mass spectra of trimethylsilyl sterols, and therefore, the authors show the main ionization reactions and the ensuing products which depend on the carbon number, the number, and position of double bonds, and the structure of the alkyl group. The authors compared retention times from two equivalent columns coupled to the most commonly used detectors: the mass spectrometer and flame ionization detector. The authors selected characteristic sterols from primary producers in tropical coral reef food webs to trace them in consumers. The phytoplankton sterol episterol was positively correlated with the triacylglycerol to sterol (TAG: ST) ratio as an indicator of good nutritional condition previously published for the same food wed. This was true principally in zooplankton, the bivalve *Pinna carnea* and top predators, such as the perciform fish *Bodianus rufus*, *Ocyurus chrysurus*, and *Caranx hippos*. In contrast, the coral *Montastrea cavernosa* had a significant correlation of storage lipids such as wax and steryl esters with the zooxanthellae sterol gorgosterol, confirming that a better nutritional condition comes from the contribution of zooxanthellae.

Chapter 2 - Sponges are ancient and simple animals with metabolic richness that has caught the attention of the scientific community. With more than 8,000 species, sponges are abundantly observed in marine, as well as freshwater, environments. Studies regarding the chemical composition and pharmacological properties studies of these organisms mainly focus on the marine species. There are four phyla, with only species of one, the Demospongiae, found in freshwater environments. The species within Demospongiae can be further divided into six living families and only fossil records for a seventh. Among the metabolites of the sponge are the sterols. These are of particular importance due to their great diversity, and thus huge biotechnological potential. Sterols have been described from different sponge families from different continents, however, few studies have been conducted involving the sterols of the freshwater sponge compared to those of the marine environment. The first report of sponge sterols dates back to 1941, which described the presence of 5,6-dihydrostigmasterol in the species *Spongilla lacustris*. Since then, it has been shown that sponges can acquire their sterols through different processes, such as absorption and modification, which highlights the influence of the medium on the diversity of these metabolites. The number of sterols present varies with the species, not being uncommon the presence of a major sterol with the others appearing in trace concentration. These characteristics make it possible to use this data as chemosystematic markers.

Chapter 3 - Olive oil is obtained from the fruit of several cultivars of olive tree (Olea europea L.), with particular characteristics. Each one of these cultivars exhibits specific physical and biochemical characteristics, providing oils with different compositions and performances.

In olive oil, sterols constitute the majority of the unsaponifiable fraction. In recent years there has been increased interest in the sterols of olive oil for their health benefits and their importance to virgin olive oil (VOO) quality. Several factors are known to affect the sterolic profiles of olive oil. Among these factors is the nature of the cultivar. Recently, it has also been proposed that these profiles could be used to classify virgin olive oils according to their fruit variety. The main sterols found in olive oils were β-sitosterol, Δ5-avenasterol, campesterol and stigmasterol. Cholesterol, 24-methylenecholesterol, clerosterol, campestanol, sitostanol, Δ7-stigmastenol, Δ5,24-stigmastadienol, and Δ7-avenasterol were also found. Most of these compounds are significantly affected by the cultivar type. Chemical characterization of monovarietal olive oil is imperative for the selection of quality cultivars that produce virgin olive oil with good quality. The sterol fraction can be considered as a useful tool to characterize and discriminate monovarietal VOOs. Thus, this chapter is devoted to recent findings concerning the effect of cultivar on sterolic profile of extra virgin olive oil.

Chapter 4 - This work was carried out on the study of virgin olive oil from two new olive varieties obtained through uncontrolled crossings. Preliminary work evaluating the oil fatty acid composition of the oil of 50 descendants showed the performance of two cultivars among the studied hybrids. These two new cultivars (B1 and B2) have an improved oil composition compared to that of Chemlali, the most dominant Tunisian olive oil variety. A further study was therefore required for their complete characterization. In the present study, the authors proposed to determine the sterolic composition. Considering the percentages of the major sterols identified and quantified in samples, β-sitosterol was the major compound for all oils with percentages of apparent β-sitosterol (sum of β-sitosterol, clerosterol, Δ-5-avenasterol) of 93%. Campesterol and stigmasterol were in all cases very close to the legal limits for olive oil ($\leq$ 4% and < campesterol, respectively). The statistical analysis showed significant differences between oil samples, and the obtained results showed that the great variability of the oil composition between varieties is influenced exclusively by the genetic factor. This study aims to contribute to the optimisation and valorisation of virgin olive oil quality in the world olive-producing areas.

Chapter 5 - Plant sterols or phytosterols are a family of phytochemicals and common components of legumes, cereals and plant oils, seeds and nuts. They are well known for their cholesterol-lowering effect in humans and have other beneficial effects on health since they inhibit colon cancer development and may prevent some cardiovascular and inflammatory diseases.

The sterolic fraction of argan oil (Argania spinosa L. skeels) and olive oil (Olea europaea L. cv. Chemlali), were investigated and compared. The total phytosterol content ranged from 1700.80mg/kg in chemlali oil to 150.40mg/kg in argan oil. In contrast to chemlali oil in which β-sitosterol is predominant, with 85.8%. The major sterols detected in the argan oils were schottenol and spinasterol. Detection of edible oil adulteration is of utmost importance to ensure product quality and customer protection. Campesterol, a sterol found in argan and Chemlali olive oil, represents less than 0.4% of total sterol content. Interestingly, argan oil contains only traces of campesterol. On the other hand, schottenol and spinasterol were not detected in chemlali oil.

Chapter 6 - Olive oils are distinct from other vegetable oils because they may be consumed without extensive refining. Virgin olive oils attract a higher price than refined olive oils because of their pleasant, rather delicate flavour and aroma and limited production volume. As a result, olive oils are subject to two types of deliberate adulteration. The first is the blending of virgin olive oils with olive oils of lower grade (e.g., refined olive oil or olive-pomace oil). The second is the less subtle mixing of olive oil with liquid vegetable oils. As a consequence of such adulteration, the International Olive Oil Council (IOOC) and the Codex Alimentarius Commission (CAC) have produced standards for virgin and refined olive and olive-pomace oils and certain blends of these products. Along with the profile of fatty acids, triglycerides and tocopherols, the sterol profile is an important parameter to assess the identity and authenticity of fats and oils; it is widely accepted as one of the most important markers for the detection of adulterated olive oils.

During the analysis of the sterolic profiles, the most frequent deviations observed in the samples of extra virgin olive oil were in: campesterol, Δ7-stigmastenol, apparent β-sitosterol, total sterols, erythrodiol and uvaol. In the case of the lampante samples, deviations were noted in the following criteria: Δ7-stigmastenol, apparent β-sitosterol, total sterols, erythrodiol + uvaol. Only two parameters deviated from the official limits in the olive pomace oils tested: Δ7-stigmastenol and apparent β-sitosterol. Campesterol and Δ7-stigmastenol were the parameters for which the most samples exhibited deviations.

These results are very important in fighting fraud and ensuring that the olive oil that consumers buy for its health or sensory properties has not been mixed with other, cheaper vegetable oils. The purpose of this paper, however, is to show how the standards of the European Union legislation for fats and oils can help to verify olive oil authenticity using the sterolic profile. This review proposes possible solutions to safeguard the consumer and protect the olive oil market.

Chapter 7 - Determination of phytosterols content, including individual components, is, at present, an inevitable part of any intensive research on plants because of their nutritional value or impact on human health. Also, the increased requirements on food quality and authenticity, as well as the expansion of investigations on health beneficial effects of phytosterols, have resulted in the searching and development of efficient analytical methods for their determination. In addition, since each plant has its specific sterol composition, phytosterols can be successfully used as markers of the authenticity of commercial edible oils and fats, easily revealing any attempt to adulterate food products of animal origin like cheese and butter, or cosmetics.

Analysis of phytosterols is a multistage procedure that includes extraction, isolation/purification as a group of related compounds and a chromatographic technique for separation, identification, and quantification (if required). The main approaches of each step that are widely used at present are presented in this chapter.

In: Sterols: Types, Classification and Structure   ISBN: 978-1-53617-231-7
Editor: Scott Jimenez   © 2020 Nova Science Publishers, Inc.

Chapter 1

IDENTIFICATION AND QUANTIFICATION OF STEROLS IN CORAL REEF FOOD WEBS

Jorge A. Del Angel-Rodríguez[1,*], PhD, Laura Carreón-Palau[1,2,†], PhD and Christopher C. Parrish[1], PhD

[1]Memorial University of Newfoundland, Ocean Science Centre, St. John's Newfoundland, Canada
[2]Centro de Investigaciones Biológicas del Noroeste, S. C. (CIBNOR), La Paz, Baja California Sur. México

ABSTRACT

This chapter describes in detail the methodology used to determine sterols in coral reef food webs and how we use the sterol data in trophic ecology studies. We briefly explain the basics of gas chromatography and mass spectrometry as necessary tools to confidently separate, identify, and quantify sterols. We describe the steps and reactions required to form sterol-trimethylsilyl ethers. This is needed to improve the

---

[*] Corresponding Author's Email: delangelj@gmail.com; cparrish@mun.ca.
[†] Corresponding Author Email: lcarreon@cibnor.mx.

chromatographic separation and identification via mass spectrometry. A main goal of this chapter is to facilitate the interpretation of mass spectra of trimethylsilyl sterols, and therefore, we show the main ionization reactions and the ensuing products which depend on the carbon number, the number, and position of double bonds, and the structure of the alkyl group. We compared retention times from two equivalent columns coupled to the most commonly used detectors: the mass spectrometer and flame ionization detector. We selected characteristic sterols from primary producers in tropical coral reef food webs to trace them in consumers. The phytoplankton sterol episterol was positively correlated with the triacylglycerol to sterol (TAG: ST) ratio as an indicator of good nutritional condition previously published for the same food wed. This was true principally in zooplankton, the bivalve *Pinna carnea* and top predators, such as the perciform fish *Bodianus rufus*, *Ocyurus chrysurus*, and *Caranx hippos*. In contrast, the coral *Montastrea cavernosa* had a significant correlation of storage lipids such as wax and steryl esters with the zooxanthellae sterol gorgosterol, confirming that a better nutritional condition comes from the contribution of zooxanthellae.

Keywords: sterols, GC-MS, FID

1. INTRODUCTION

Lipids can be simple or complex. Simple lipids are compounds that produce no more than two primary products per molecule during hydrolysis. They usually constitute energy reserves in aquatic food webs as they contain fatty acids (FA) linked by an ester bond to either monoalcohols as in waxes (WAX), or a polyalcohol (glycerol) as in triacylglycerols (TAG), diacylglycerols (DAG) and monoacylglycerols (MAG). Some simple lipids such as sterols (ST), which are ring-chained alcohols, perform structural functions in cell membranes (Parrish et al., 2000).

In contrast, complex lipids are compounds that produce three or more primary products per molecule during hydrolysis. They usually form the structure of cell membranes and include the phospholipids or glycerophospholipids, consisting of a molecule of glycerol linked to one or two fatty acids and to a polar phosphate group, which is in turn linked to another group such as an amine. Other complex lipids are the glyceroglycolipids: a glycerol linked to one or two fatty acids and to a mono- or disaccharide, and the sphingolipids: fatty acids linked by an amide bond to long-chain bases (or sphingoid bases) and occasionally to other amines (Christie, 2003).

According to Christie (2003), other biomolecules should also be classified as lipids. For instance, fatty acid derivates such as esters and amides; substances related biosynthetically to fatty acids such as prostanoids and aliphatic ethers or alcohols; and substances related functionally to fatty acids such as sterols and even bile acids.

Despite their breadth, those criteria still exclude some compounds with little functionality or structural relationship with the previous ones. Steroids, fat-soluble vitamins, pigments, carotenoids, and terpenes are soluble in organic solvents such as chloroform, benzene, ethers and alcohols and therefore can be extracted and quantified together with other lipids. Accordingly, a careful separation of the different lipid compounds allows a better interpretation of lipid processes under investigation.

1.1. Sterols and Stanols

Sterols, also known as steroid alcohols, are tetracyclic compounds. This subgroup of steroids has a hydroxyl group (alcohol-OH, hence, the "ol" ending) at position 3 of the first ring (A-ring). They are amphipathic lipids as the hydroxyl group in the A-ring is polar, and the rest of the aliphatic chain is non-polar. They are synthesized from acetyl-coenzyme A via the 3-hydroxy-3-methyl-glutaryl-coenzyme A reductase (HMG-CoA reductase) route. Contrary to sterols, stanols do not have a double bond in the ring structure.

Sterols and related compounds are a class of organic molecules that play essential roles in the physiology of eukaryote organisms. For example, cholesterol is part of the cell membrane in animals affecting the fluidity of the membrane and it serves as a second messenger in developmental signalling. In animals, corticosteroids such as cortisol act as signals in cell communication and metabolism in general. Sterols are a common component of human skin oils (Christie and Xianlin, 2012).

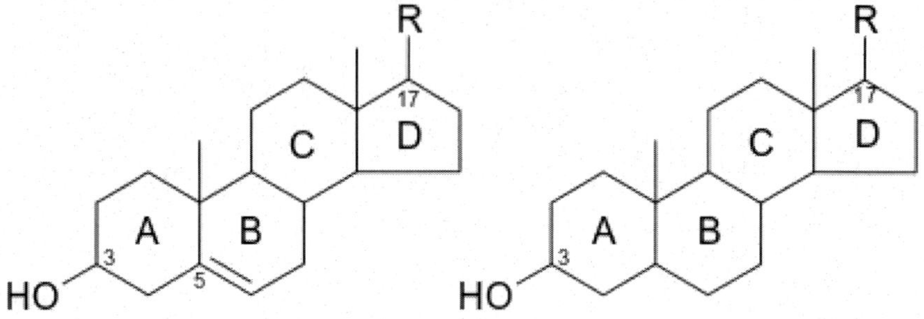

Figure 1. General structure of a sterol (left) and a stanol (right) and ring notation.

Photosynthetic organisms usually have different sterols, such as stigmasterol, sitosterol, ergosterol, brassicasterol, etc. Depending on the species, the relative quantities and proportions of different sterols may vary, hence their use as biomarkers. For example, dinosterol and dehydrodinosterol have been reported in dinoflagellates, while fucosterol and fucostanol have been observed in kelp, brassicasterol, and desmosterol in diatoms.

Cholesterol is usually a minor component in plant cells; however, it is the principal sterol in red algae. Sterol esters (sterols linked to a fatty acid) are also present in plant cells, but in general, they are relatively minor components (Parrish et al. 2000). In coral-reef food webs, sterol biomarkers have been defined as primary-producer sterols that are distinctively detected in primary consumers. For a tropical coral-reef food web the most notable are: *trans*-22-dehydrocholesterol (sea grass), stellasterol (green algae), brassicasterol (red algae), poriferasterol (brown algae), episterol[2] (phytoplankton), and gorgosterol (zooxanthellae) (Carreón-Palau, 2015).

---

[2] Jones et al., (1994) referred to 24-methyl-5α-cholesta-7,24(28)-dien-3β-ol as 24-methylenelophenol, and Carreón-Palau (2015) used the same common name. However the structure, molecular weight and ions reported by both authors matches that of episterol in PubChem and lacks the methyl group on C_4 described for 24-methylenelophenol.

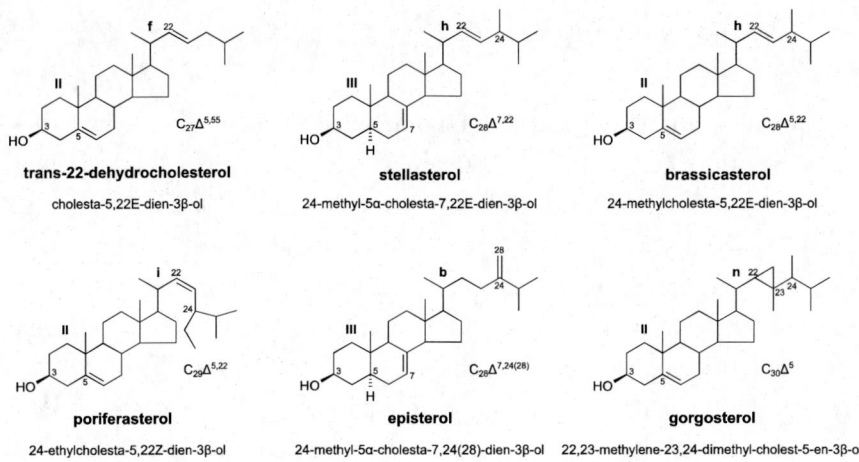

Figure 2. Phytosterols found in primary producers of tropical coral reefs. Common names, IUPAC names and shorthand notation. Bold roman numerals for rings, and letters for sidechains, follow Jones et al. (1994) notation.

Cholesterol, the most common sterol found in animals, is vital for membrane structure of the animal cell and works as a precursor for fat-soluble vitamins and steroid hormones. In fungi, ergosterol is present in cell membranes and has a similar function to animal-cell cholesterol. Plant sterols are named phytosterols, and whereas they appear to be similar to cholesterol, they differ in the side chain or the presence/absence of double bonds.

Phytosterols extracted from oils are insoluble in water, relatively oil-soluble and alcohol soluble. Campesterol, sitosterol, and stigmasterol are among the most common phytosterols. The most common stanols in the human diet are sitostanol and campestanol (Christie and Xianlin, 2012).

Clinical trials have demonstrated that phytosterols block absorption sites for cholesterol in the human intestine, hence their use as food additives. Still, little investigation in collateral effects makes them only recommendable for people with high cholesterol levels, but not for pregnant or lactating women (García-Llatas y Rodríguez-Estrada, 2011). Preliminary research suggests that phytosterols can decrease cancer incidences as well (Ramprasath and Awad, 2015).

2. STEROLS ANALYSIS BY GAS CHROMATOGRAPHY

Chromatography is a method for chemical separation in which the components distribute between a stationary phase and a mobile one that moves in a defined direction. In gas chromatography (GC) the mobile phase is a gas, and if the stationary phase is liquid then it is called gas-liquid chromatography (GLC) (McNair and Miller, 1997). The carrier gas should be chemically inert, e.g., helium, argon, nitrogen, or hydrogen according to the type of detector used.

For over half a century, gas chromatography coupled to flame-ionization detectors (GC-FID) and mass-selection detectors (GC-MS) have been the method of choice to analyze thermostable molecules. Biomolecules that can be analyzed via GC-FID or GC-MS include fatty acids, hydrocarbons, sterols, alcohols, and amino acids among others.

2.1. Identification and Quantification of Coral Reef Sterols by GC-FID and GC-MS

GC-FID is considered the best option to do quantitative evaluations of sterols because all the compounds are ionized to CO_2 and the compound amount can be directly related to the area under the chromatographic peak. However, there are plenty of sterols eluting at similar retention times, and they can be misidentified. In contrast, mass spectrometer detector (MSD) allows a better identification because ionization is incomplete resulting in characteristic ions along with the molecular ion. For instance, with GC-FID, lathosterol elutes at 51.53 minutes, while stigmasterol elutes at 51.18 minutes, and they could be confused. However, molecular ions obtained with MSD at retention times of 48.18 and 48.24 are 458 and 484, respectively (Table 2).

Sterols can be extracted from the biological matrix following the same procedure used for other lipids. If triacylglycerols (TAG) are being transesterified to obtain fatty acid methyl esters (FAME), then FAME will be re-suspended in hexane for injection in GC-MS or GC-FID. The remaining hexane can be dried under a gentle flux of nitrogen gas, and 100 µL of N,O-bis (trimethylsilyl) trifluoroacetamide (BSTFA) (CAS: 25561-30-2) is added. Then the mixture is heated at 85°C for 15 minutes in an oven. The sample is further dried with nitrogen gas and hexane is added to achieve a lipid concentration close to 1000 µg mL^{-1}. If lipid classes were previously determined (e.g., by Chromarod-Iatroscan TLC-FID), then the total amount of sterols in the sample will be known. This will allow preparation of a hexane solution where each sterol ranges from 20–200 µg mL^{-1}.

For sterol separation, a 0.25 mm x 0.25 µm DB-5 MS column can be used. The recommended length is 60 m, but 30 m will suffice. The carrier gas will be helium (He), and recommended chromatographic conditions will be as follows: Helium flow 1.2 ml min^{-1}, injector temperature 250°C. Immediately after injection, the column temperature (oven) will remain at 60°C during 1 min, and then it will increase to 100°C at a rate of 25°C min^{-1}. After reaching 100°C, the temperature will increase to 150°C at a rate of 15°C min^{-1}. The last ramp consists of an increase in temperature until 315°C at a rate of 3°C min^{-1}. The quadrupole (MS-Detector) is set to 150°C, and the ion source of the mass spectrometer (MS) is set to 270°C and 70 eV. Sterol identification is based on mass interpretation of the observed ions as described by Jones et al. (1994) using Wsearch 32 free-distribution software (Table 2.).

Once identified, retention times of the gas chromatograph coupled to a mass spectrometer (GC-MS) can be compared against the retention times obtained with a gas chromatograph coupled to flame ionization detector (GC-FID). For this latter case, a similar column should be used such as the DB-5 30 m x 0.32 mm x 0.25 µm, following the same chromatographic method (Table 2.).

2.2. Injection Systems

Micro-syringes are the most common method to inject a liquid or gas sample (~1 µL). The injection occurs through a silicone septum into a vaporization chamber (injection port) at the head of the column. The temperature of the chamber must guarantee the volatilization of the different compounds within the sample in order to ensure their further separation (McNair, 1981; Christian, 1981; Lederer and Lederer, 1957). Most injection systems control the amount of sample entering the column; thus a split injection will discard a pre-establish amount of sample before volatilization, whereas a splitless injection will inject the whole sample into the volatilization chamber. A split-splitless injection will allow discarding a controlled small amount of sample during a few seconds. Finally, if a sample is injected as a gas, an on-column injection will allow a direct injection into the column bypassing the injection port.

2.3. Column

The widely used capillary columns allow for small sample sizes and, as long as the carrier gas is hydrogen, helium or nitrogen, their response sensitivity is high (Ackman, 2002). The temperature of the column-containing oven can modify the column efficiency and is an important factor for good resolution of compounds. Optimum column temperature depends on the physicochemical features and components of the mixture and upon the selectivity capacity of the stationary phase.

2.4. Flame Ionization Detector (FID)

In this detector, the analyte in the effluent (carrier gas) enters the FID and passes through the air-hydrogen flame causing ion formation and originating an electric current between two electrodes (cathode and anode). FIDs have a wider linear range and are considered universal detectors (Kitson et al., 1996). Analytes with higher carbon number produce a higher signal.

2.5. Mass Selective Detector (MSD)

The main purpose of the MS is to transform the analyte into measurable products that indicate the nature of the original molecule. Such products are ions, usually, with a positive charge in the gas state, which can be separated according to their molecular mass. The mass spectrum is a plot of the mass-charge ratio from each ion and its relative abundance. The abscissa indicates the mass-charge (m/z) ratio of each of the ions that form the molecule and the ordinate is the relative abundance of each ion (McLafferty and Tureček, 1993).

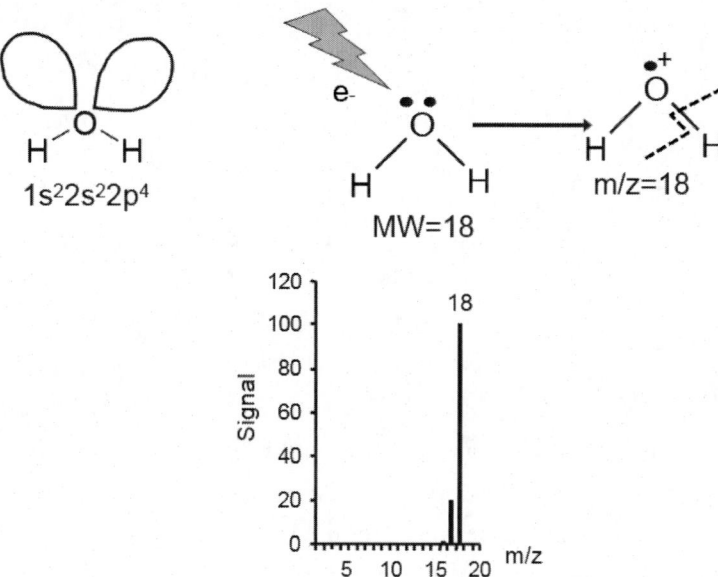

Figure 3. Ionization of the water molecule.

2.5.1. Theory of Unimolecular Ion Degradation

To understand the strengths and limitations of mass spectrometry, it is important to review the basic theoretical aspects of unimolecular ion decomposition. For gas ions, only unimolecular reactions (one ion per molecule) can be generated from the sample molecule (). The probability of each ion being formed will depend on the molecular structure and its internal energy. As there is a population of molecules, there will be different proportions of ions. The ion that requires less energy to be ionized is the whole molecule (molecular ion or M⁺), and its presence and relative contribution are key to identify the compound in their mass spectra (McLafferty and Tureček, 1993).

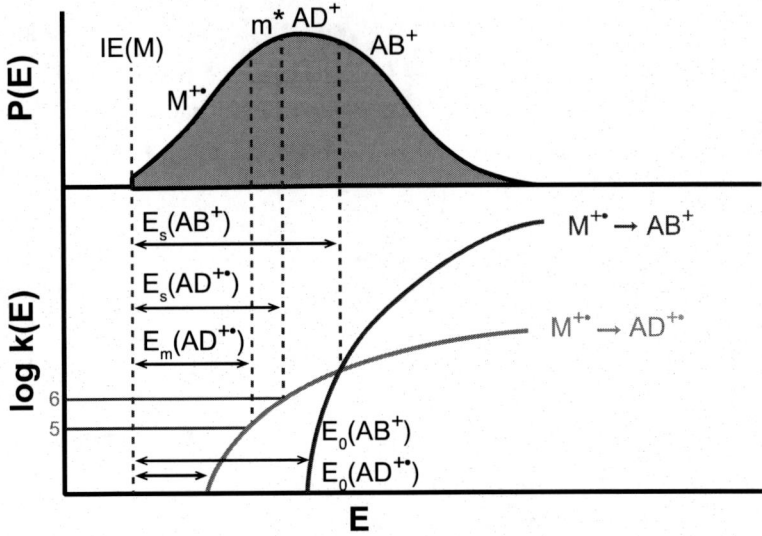

Figure 4. Wahrhaftig's diagram showing the probability of formation (P) of the molecular ion or its products as a function of the internal energy (E) of the molecule (Modified from McLafferty & Tureček, 1993).

2.5.2. Description and Functioning of a Mass Spectrometry Detector

The mass spectrometer is a type of detector that ionizes molecules. The stream of a vaporized sample enters the ionization chamber from the GC-MS interface with an approximated pressure of 10^{-5} Torr. There, it is hit by the electron current created by an incandescent filament and guided by a magnetic field. The high-energy electrons interact with the sample molecules, ionizing and fragmenting them. The positive voltage on the repeller pushes the positive ions into several electrostatic lenses. These lenses concentrate the ions into a tight beam, which is directed towards the analyzer. There, a quadrupole filter separates them according to their mass:charge (m/z) ratio.

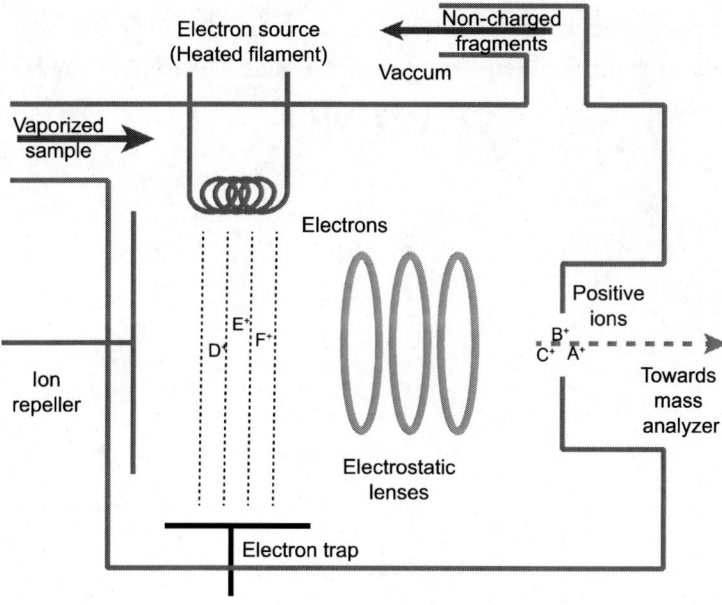

Figure 5. Schematic representation of the ionization chamber (Modified from Agilent Technologies, 2012).

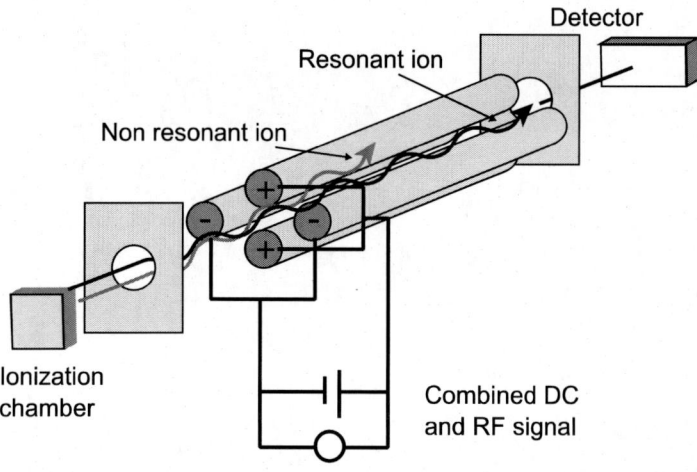

Figure 6. A quadrupole filter mass analyzer (Modified from McLafferty and Tureček, 1993).

A vacuum pump accelerates the ions towards the detector where the ion current is amplified generating a signal that is interpreted by software and is represented by a mass spectrum (McLafferty and Tureček, 1993, Siuzdak, 1996, Gunstone, 1996).

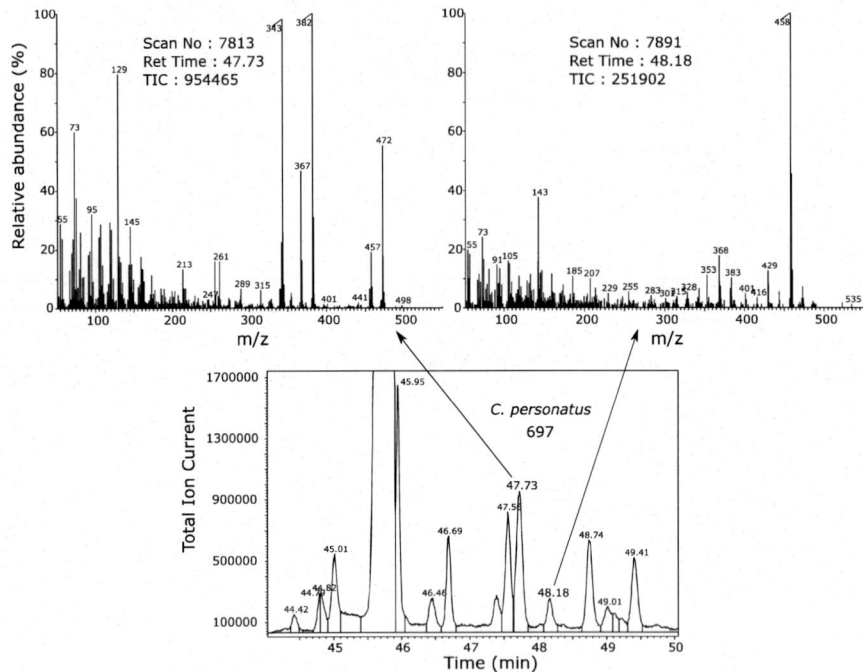

Figure 7. Mass spectrum of two adjacent chromatographic peaks from a reef fish depicting the relative abundance (%) of different m/z ions at a particular scan, and retention time. Peak height in the chromatogram is the total ion current (TIC) at that particular scan.

2.5.3. Ionization Reaction of Sterol-Trimethylsilyl Ethers

Sterols are better analyzed separately from other compounds in non-polar phases either in a free form or as sterol-trimethylsilyl ethers (sterol-TMS ethers), where a silicon atom bound to three methyl groups $[Si(CH_3)_3]$ substitutes for the hydrogen in the alcohol group (OH) of the sterol.

Shows the designation of carbon atoms in cholesterol-TMS ether. Such a numbering system is used both in formulae as well as in ionization reactions. It is important to note that similar to fatty acid methyl esters (FAME), the carbons of the methyl groups of sterol-TMS ethers are not included in the numbering, or in the shorthand notation, but they should be considered for calculating the molecular weight for MS interpretation.

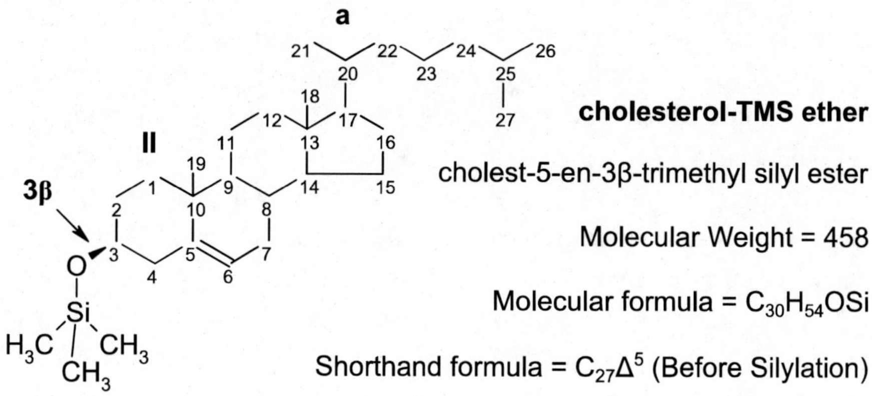

cholesterol-TMS ether

cholest-5-en-3β-trimethyl silyl ester

Molecular Weight = 458

Molecular formula = $C_{30}H_{54}OSi$

Shorthand formula = $C_{27}\Delta^5$ (Before Silylation)

Figure 8. Scheme of cholesterol-TMS ether, IUPAC name, molecular weight, and shorthand notation.

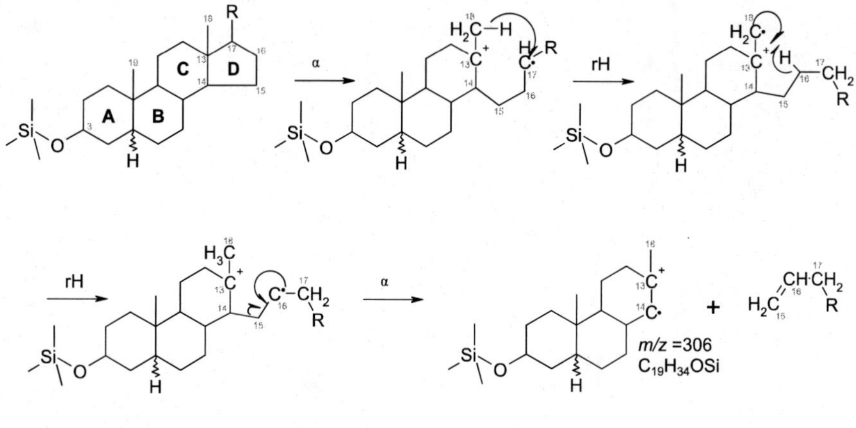

Figure 9. Production of carbenium ions and reciprocal transfer of hydrogen with cleavage of C_{14} and C_{15} in a stanol-TMS ether (modified from McLafferty and Tureček, 1993).

During ionization, fragmentation of the D-ring and loss of the R-chain (starting from C_{20}) are characteristic of C_{17}-substituted steroids, but they do not depend on the presence and nature of the substituents on the A and B rings. Accordingly, although the following reactions were originally described by McLafferty and Tureček (1993) for steroids (compounds without the -OH at C_3), they also apply to sterols and their TMS ether derivatives.

One of the most common ions obtained during ionization is the (M-CH_3)$^+$, that comes from the loss of angular methyl groups (C_{18} and mainly C_{19}). A favored reaction is the cleavage between C_{13} and C_{17} (on the D-ring) producing a stable carbenium ion. Reactions at the site of the radical of this species can proceed by two main pathways: the first one is the reciprocal transfer of hydrogen and cleavage between C_{14} and C_{15} resulting in an ion of m/z=306 for stanol-TMS ethers.

The second pathway is electron donation to the neighboring bond. This will produce a cleavage between C_{15} and C_{16} and the loss of the substituent group at C_{17} together with carbons C_{16} and C_{17}. As a consequence, this will result in an ion with m/z=320. In this route, further loss of a methyl group at C_{19} from the ion with m/z=320 contributes substantially to produce an ion with m/z=305.

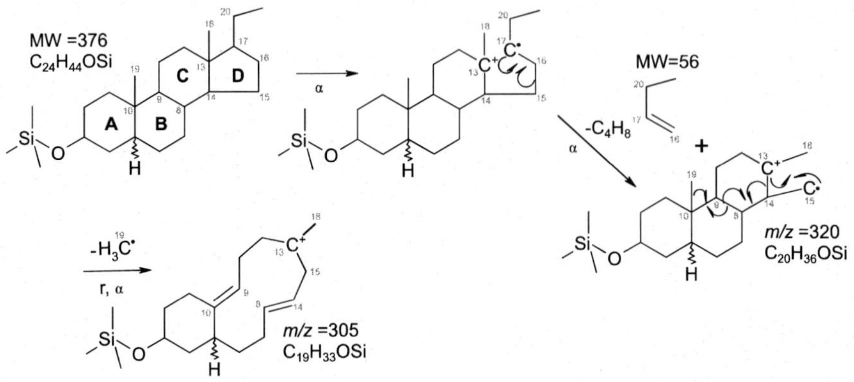

Figure 10. Production of carbenium ions and electron donation to the neighboring bond with cleavage of C_{15} and C_{16} in a stanol-TMS ether (modified from McLafferty and Tureček, 1993).

Finally, the ion with m/z=237 is composed mainly of the A and B rings but is formed by a complex mechanism in conjunction with the triple transfer of hydrogen. Despite the apparent degree of randomness, stereochemistry of A and B rings has a characteristic effect in the process: in the mass spectrum of the isomer 5-β-(cis), the ratio [m/z 237]/[m/z 239] is much lower than that of the isomer 5-α. This effect is relatively independent of the substituent group at C17.

In most sterol-TMS ethers, the molecular ion (M⁺) is preserved after ionization. This allows both the carbon number and the number of double bonds to be inferred. Those details are particularly useful in molecular identification (Table 1). For instance, cholesterol-TMS ether (MW=458) will have 27 carbons and one double bond at carbon 5 ($27\Delta^5$).

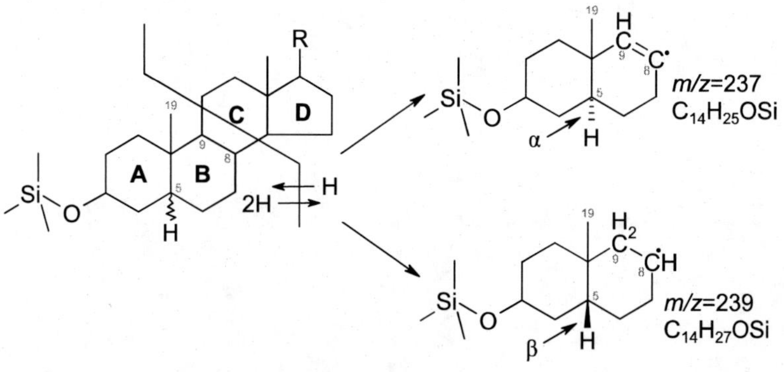

Figure 11. Cleavage of A and B rings from the main molecule of a stanol-TMS ether (modified from McLaferty & Tureček, 1993).

Table 1. Molecular weight of sterol-TMS ethers as a function of the number of carbon atoms (#C) and double bonds (#DB) in the molecule

#DB	Number of carbon atoms in sterol (#C)				
	26	27	28	29	30
0	446	460	474	488	502
1	444	458	472	486	500
2	442	456	470	484	498
3	440	454	468	482	496
4	438	452	466	480	494

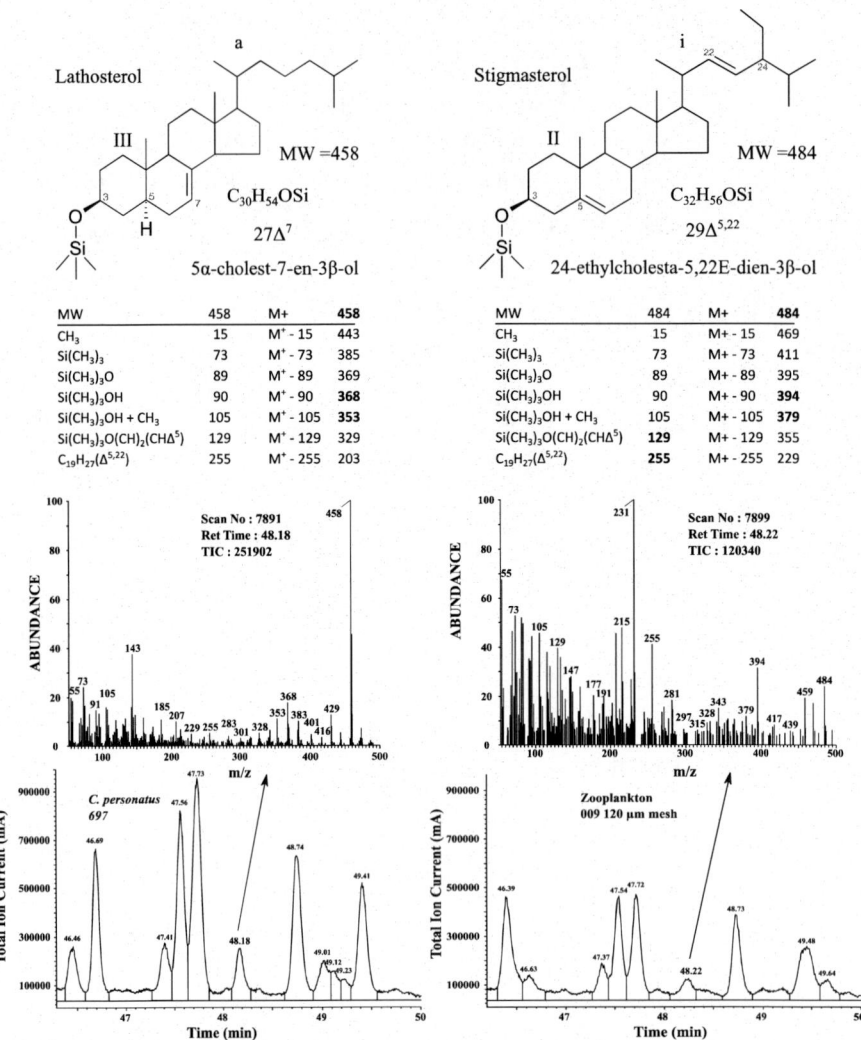

Figure 12. Example of two coral reef sterols with similar retention times but differences in molecular structure, characteristic ions and mass spectra. Tables show the ions expected depending on structure. Bold ions in tables are clearly visible in the mass spectra and are confirmatory of structure.

Furthermore, characteristic ions can be inferred by subtraction of the different fragments from the molecular ion. Some important fragments for interpretation of the whole structure are (M-R)$^+$, (M-R-2H)$^+$, (M-H2O)$^+$, (M-R-H2O)$^+$, (M-R-2H-H2O)$^+$, and (M-CH3-H2O)$^+$. Other fragments will indicate the position of double bonds. For instance, lathosterol-TMS ether will have a MW=458. This means the molecule has 27 carbons and one double bond. We should expect a signal at m/z=213 indicating a double bond at C_7. For stigmasterol-TMS ether, double bonds at C_5 and C_{22} are shown by ions with m/z=129 and m/z=255.

2.5.4. Software-Assisted Interpretation of Mass Spectra

Most GC-MS instruments have specialized software for processing mass spectra and allow a certain level of interpretation. Unfortunately, each company works with different file formatting, and sometimes they are incompatible. Furthermore, in most cases, software licence numbers are restricted, making it illegal to install on personal computers, limiting the time a person can devote to analyze chromatograms.

However, there are free alternative packages fully capable of reading any format file. The most well-known is the package created by the National Institute of Standards and Technology (NIST) that comprises 2 main programs: 1) "AMDIS": a package that extracts the mass spectra from the individual peaks from a GC/MS file, and 2) "MS Search Program": a package that helps searching mass spectra on a database of known substances. There are also commercial databases from known spectra that can be acquired from distributors. Both NIST packages can be downloaded at: http://chemdata.nist.gov/mass-spc/ms-search/.

2.6. Quantification of Sterols by GC-MS and GC-FID

The area under the peak in GC-MSD is the sum of the total ion current. Therefore, peak area results from the combination between number and mass of ions, and ion frequencies, and it can change among sterols. Peak area units are equal to milli Ampers per minute (mA × min). However, most packages convert minutes into seconds. Accordingly, area can also be expressed as mA × sec.

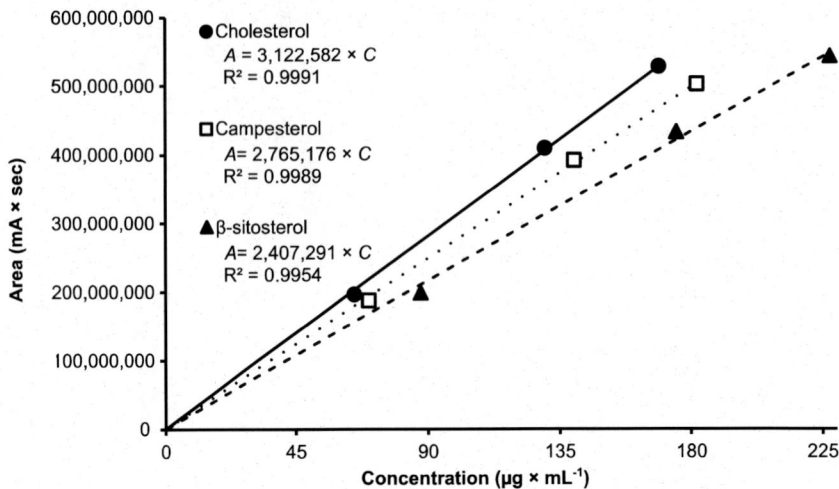

Figure 13. Calibration curve for cholesterol, campesterol and β-sitosterol representing the area (A) under the peak (mA × sec) as a function of concentration C (µg mL^{-1}). Sample concentration is equal to the area divided by the slope. See text for procedures and unit explanation.

Differences in MS-detector response (slope) among sterols are calculated by plotting the areas (mA × sec) of integrated chromatographic peaks against known concentrations of commercial standards (µg mL^{-1}). A calibration curve should be created with commercial standards of sterol-TMS ethers with different molecular weights, such as cholesterol (458), campesterol (472) and β-sitosterol (486).

Concentrations of standards should be within a range where signal:noise ratio >10, and where area and concentration correlate linearly (~10–250 µg mL^{-1}). Linear regression analysis on each plot yields the slope for each sterol ([mA sec]/ [µg mL^{-1}]) with $R^2 > 0.99$. Then, volumes of hexane of samples are adjusted in order to obtain a concentration of each sterol in the linear range.

Any software capable of peak integration can be used to quantify sterols. Under the described chromatographic conditions, sterols elute at the end of the chromatogram after minute 40 (Section 0), and once identified, they should be integrated to generate a list of absolute areas.

Quantification (µg mL^{-1} of hexane) of samples is achieved by dividing the areas from integrated chromatographic peaks by their respective slope.

This concentration is then multiplied by the volume of hexane (mL) added to each sample and divided by the amount of extracted sample dry or wet biomass (mg). The sterol concentration is then reported per dry biomass or wet biomass ($\mu g\ mg^{-1}$ or $mg\ g^{-1}$), or as percentages of the total amount of sterols.

In contrast, in GC-FID, areas are directly summed and sterol proportions are the proportional part of the sum of all areas. This proportion can be further related to total sterols computed by the concentration of steryl ester plus free sterol fractions of dry biomass. Those can be obtained by Iatroscan-FID, and an approximation of individual sterol concentrations per unit of dry or wet biomass is obtained ($\mu g\ mg^{-1}$ or $mg\ g^{-1}$). If Iatroscan information is not available, a calibration curve should be used as described before. Sample analysis by GC-MSD is more expensive than GC-FID, therefore one sample can be used for identification, and the replicates can be injected in GC-FID for quantification. A comparison between retention times on both GC-MSD and GC-FID is presented in Table 2.

Table 2. Ions in sterols. RT= Retention time, MSD=Mass Spectrometry Detector, FID= Flame Ionization Detector, MW=Molecular Weight. Source: Carreón-Palau, 2015

RT GC-MS	RT GC-FID	Compound	MW	Key ions for identification	Structure
41.86	44.54	24-nordehydrocholesterol	442	69, 97, 211, 353, 366, 442	24-norcholesta-5, 22E-dien-3β-ol
43.95	46.92	coprostanol	460	75, 215, 257, 355, 370, 403, 445, 460	5β-cholestan-3β-ol
44.40	47.37	24-Nor-22, 23 methylenecholest-5-en-3-β-ol	456	55, 69, 111, 129, 327, 366, 456	24-Nor-22, 23 methylenecholest-5-en-3β-ol
44.81	47.10	occelasterol	456	55, 69, 111, 129, 255, 327, 351, 366, 441, 456	27-nor-24-methylcholest-5, 22E-dien-3β-ol
45.01	48.23	trans 22 dehydrocholestanol	458	73, 91, 345, 374, 458	5α-cholest-22E-en-3β-ol

Table 2. (Continued)

RT GC-MS	RT GC-FID	Compound	MW	Key ions for identification	Structure
45.01	48.25	patinosterol	458	69, 81, 107, 257, 345, 374, 389, 404, 443, 458	27-nor-24-methyl-5α-cholest-22E-en-3β-ol
45.60	48.28	cholesterol	458	73, 129, 329, 353, 368, 443, 458	cholesta-5-en-3β-ol
45.75	48.44	cholestanol	460	75, 215, 305, 355, 370, 403, 445, 460	5α-cholestan-3β-ol
45.98	49.39	desmosterol	456	69, 129, 253, 327, 343, 366, 441, 456	cholesta-5,24-dien-3β-ol
45.99	49.28	7-dehydrocholesterol	456	75, 182, 351, 368, 386, 441, 456	cholesta-5,7-dien-3β-ol
46.39	49.12	brassicasterol	470	69, 129, 255, 340, 365, 380, 455, 470	24-methylcholesta-5,22E-dien-3β-ol
46.45	50.13	brassicastanol	472	75, 109, 257, 345, 374, 458, 472	24-methyl-5α-cholest-22E-en-3β-ol
46.64	49.29	ergost-7-enol	472	213, 229, 255, 343, 367, 378, 457, 472	24-methyl-5α-cholest-7-en-3β-ol
46.84	50.31	ergosterol	468	69, 75, 131, 253, 327, 342, 363, 378, 468	24-methylcholesta-5,7,22E-trien-3β-ol
47.39	50.05	stellasterol	470	73, 129, 213, 229, 255, 343, 455, 470	cholesta-7, 22E-dien-3β-ol
47.40	50.00	dihydrobrassicasterol	472	75, 209, 267, 343, 386, 472	24b-methylcolest-5-en-3β-ol
47.45	50.10	ergost-8(14)-enol	470	75, 91, 229, 255, 343, 378, 457, 470	24-methyl-5α-cholest-24(28)-en-3β-ol
47.54	50.19	24-methylenecholesterol	470	253, 257, 296, 341, 365, 386, 455, 470	24-methylcholesta-5, 24(28)-dien-3β-ol
47.72	50.37	campesterol	472	73, 129, 255, 343, 367, 382, 457, 472	24α-methylcholest-5-en-3β-ol
48.18	51.53	lathosterol	458	73, 255, 303, 353, 443, 458	5α Cholest-7-en-3β-ol
48.24	51.18	stigmasterol	484	83, 129, 255, 355, 379, 394, 484	24-ethylcholesta-5, 22E-dien-3β-ol
48.34	51.34	23-24 dimethylcholest 5, 7-dien-3 β-ol	486	98, 129, 283, 342, 381, 433, 486	23-24 dimethylcholest 5, 7-dien-3β-ol
48.73	51.44	episterol	470	75, 131, 213, 253, 365, 386, 455, 470	24-methyl-5α-cholesta-7, 24(28)-dien-3β-ol

RT GC-MS	RT GC-FID	Compound	MW	Key ions for identification	Structure
49.05	52.49	4-24 dimethyl 5, 7-dien-3-β-ol	484	69, 129, 283, 343, 355, 379, 394, 469, 484	4-24 dimethyl 5, 7-dien-3β-ol
49.14	52.61	poriferasterol	484	69, 129, 283, 343, 355, 379, 394, 469	24-ethylcholesta-5, 22Z-dien-3β-ol
49.14	52.02	spinasterol	484	55, 73, 129, 213, 255, 343, 372, 469, 484	24-ethyl-5α-cholesta-7, 22E-dien-3β-ol
49.49	52.04	β-Sitosterol	486	73, 129, 255, 357, 381, 396, 471, 486	24-ethylcholesta-5-en-3β-ol
49.42	52.94	fucosterol	484	55, 73, 129, 257, 281, 296, 355, 371, 386,	24-ethylcholesta-5,24(28)E-dien-3β-ol
49.61	52.22	sitostanol	488	75, 218, 229, 358, 381, 398, 488	24-ethyl-5α-cholestan-3β-ol
49.76	53.20	unidentified isomer of fucosterol	484	73, 129, 281, 296, 386, 469, 484	24-ethylcholesta-5,24(28)Z-dien-3β-ol
49.78	53.16	4, 23, 24 trimethyl 5α cholestenol	500	57, 283, 297, 359, 373, 387, 485, 500	4, 23, 24 trimethyl-5α-cholest-24(28)-en-3β-ol
49.87	52.55	Dinosterol	500	69, 359, 457, 500	4,23,24-trimethyl-5α-cholest-22E-en-3β-ol
50.15	52.87	Dinostanol	502	57, 73, 487, 397, 502	4,23,24-trimethyl-5α-cholestan-3β-ol
50.80	53.68	4, 23, 24 trimethylcholestenol	500	143, 227, 243, 269, 395, 410, 485, 500	4, 23, 24 trimmethyl-5α-cholest-7-en-3β-ol
50.23	53.74	Isofucosterol	484	55, 73, 129, 257, 281, 296, 386, 469, 484	24-ethylcholesta-5,24(28)Z-dien-3β-ol
50.50	53.82	Cycloartenol	498	73, 95, 109, 189, 369, 393, 483, 498	9,19-Cyclo-24-lanosten-3β-ol
51.86	55.36	Gorgosterol	498	73, 129, 255, 343, 386, 408, 483, 498	Gorgost-5-en-3β-ol

3. TRACKING STEROLS IN A CORAL REEF FOOD WEB

3.1. Sterols as Food Web Tracers

Cholesterol is the main sterol in most animals, while phytosterols are synthesized by algae and plants. Phytosterols are structurally similar to cholesterol, although they typically differ from cholesterol in the number and position of double bonds in the tetracyclic sterol nucleus or in the side chain, or by having an additional substituent, such as methyl or ethyl groups at C_{-24} in the side chain (Moreau et al., 2002).

Herbivorous insects and crustaceans use dietary sterols to synthesize cholesterol. Most studied species are capable of dealkylation and reduce common C_{24} alkyl phytosterols, such as sitosterol or stigmasterol, to cholesterol (Martin-Creuzburg and von Elert, 2009).

However, more than 200 different types of phytosterols have been reported in plant material, and not all of them are suitable cholesterol precursors.

For instance, certain phytosterols are more efficient in supporting somatic growth of the crustacean *Daphnia magna* than cholesterol (e.g., fucosterol, brassicasterol) while others are less efficient (e.g., dihydrocholesterol, lathosterol), indicating strong differences in the efficiencies with which these dietary sterols were assimilated and further processed within the body (Martin-Creuzburg et al., 2014). Therefore, phytosterols detected in consumers can be used as biomarkers as they are not easily bioconverted.

Although fish, crustaceans, and cephalopod molluscs contain cholesterol as their predominant sterol (often > 93% of sterols), fauna from other phyla contain very complex mixtures, often with twenty or more different sterols.

In some taxa, such as bivalve molluscs or sponges, a sterol other than cholesterol may be the major constituent (Goad, 1981). The sterol composition of an animal depends on three main factors. a) the spectrum of sterols encountered in the diet of the animal, b) the selectivity which the animal displays for the absorption, or excretion, of any particular compound in the mixture; and c) the assimilation by a host animal of sterols produced by symbiotic algae or associated bacteria or fungi in the digestive tract (Goad, 1981). Sterols are particularly well suited as biomarkers for nutrient acquisition in heterotrophic corals because their origin can be identified (Crandall et al., 2016).

3.2. Study Site

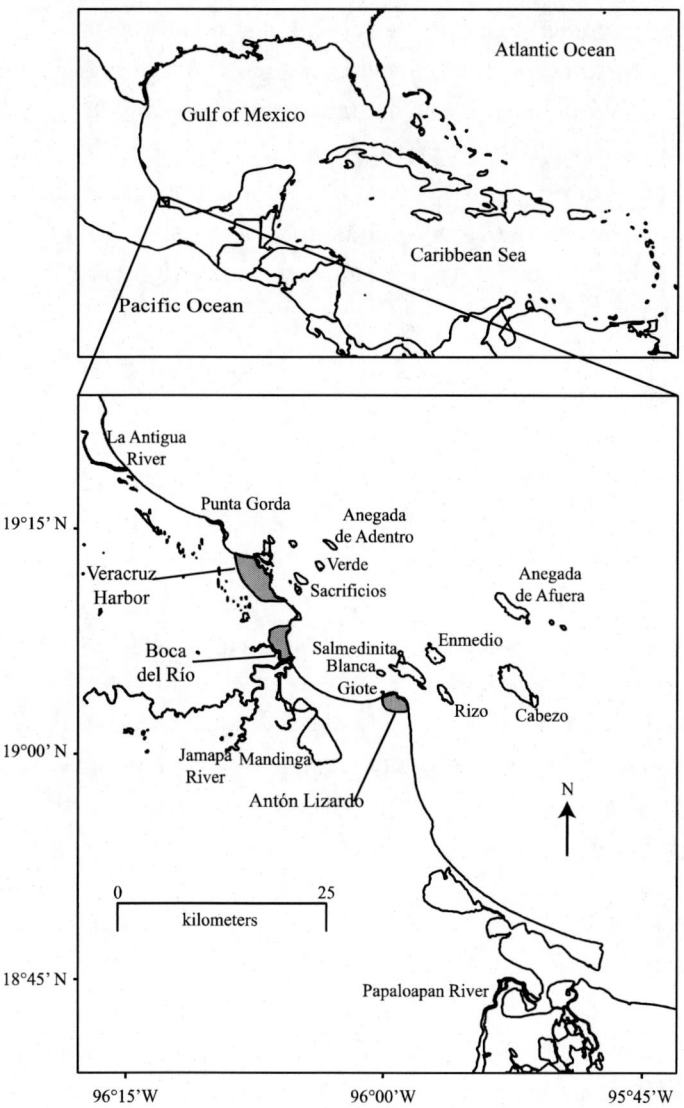

Figure 14. Study area.

The Veracruz reef system national park (Parque Nacional Sistema Arrecifal Veracruzano; PNSAV) is located off the State of Veracruz, Mexico, adjacent to the cities of Boca del Rio and Anton Lizardo

(19°02′24″ to 19°15′27″ N, 96°12′01″ to 95°46′46″ W). It is part of a larger coral reef system in the Caribbean and Gulf of Mexico. A group of 13 reefs is located near Veracruz and Boca del Rio, and another group of 15 reefs with larger structures is located near Antón Lizardo; the two reefs are separated by the Río Jamapa, and delimited on the north by Río La Antigua and on the south by the Río Papaloapan. Under sustained southward winds, a cyclonic eddy develops off Veracruz Port, which enhances productivity in the area. Under northward wind conditions or variable southward winds, the relatively high productivity area vanishes (Salas-Monreal et al., 2009).

3.3. Sampling Methods

Sampling was conducted in April–May (dry season) from 2007 to 2009. The most abundant primary producers and consumers were selected based on ecological studies (Carreón-Palau et al., 2013). We collected samples from 18 species. Plankton and multiple trophic levels up to the crevalle jack *Caranx hippos* were sampled at nine stations within the PNSAV. Those stations were: the mouth of the Río Jamapa, six reefs named Enmedio, Cabezo, Anegada de Afuera, Anegada de Adentro, Verde, and Sacrificios, one offshore station between Anegada de Adentro and Anegada de Afuera, and one offshore station near Anton Lizardo village (Carreón-Palau et al., 2013, and Carreón-Palau et al., 2018).

3.4. Data Analysis

Sterol biomarkers were defined as the distinctive sterol for each primary producer assimilated by a primary consumer or by a host, and used to detect the key sources for suspension feeders, including coral, clams, and sponges, as well as grazers, and active predators. Species differences of individual (univariate) fatty acids and sterols were tested independently in primary producers, invertebrates, and fish with one-way ANOVA using the Fisher statistic (F); residual analyses were used to test assumptions of normality and equal variance. When assumptions were not met, nonparametric Kruskal–Wallis statistic (H) was used ($\alpha = 0.05$).

Correlation between sterol biomarkers with previously published data of storage or structural lipids (Carreón-Palau et al., 2018) with the same species was assessed with the Pearson correlation statistic, using the concentration per dry weight ($\mu g\ mg^{-1}$). Significant positive correlations with storage lipids (TAG, wax and steryl esters, and TAG: ST ratio) imply primary producer(s) that promote a better nutritional condition. Significant positive correlations with structural lipids (sterol and phospholipids) imply direct incorporation of biomarkers into the cellular membrane, while significant positive correlations with acetone-mobile polar lipids imply biomarkers are coming from the thylakoid membrane of primary producers. Pearson correlation analyses were performed with Minitab 15.1.1.0 software.

3.5. Results

3.5.1. Identifying Sterol Biomarkers for Primary Producers in Coral Reef Food Webs

Sterols identified in the coral reef food web were 37 (Table 2), while primary producers had 29, ranging from four in the mangrove to 19 in phytoplankton (Table 3Table 3). Primary producer sterols such as *β*-sitosterol were the main sterol in the mangrove *R. mangle* at 82±1%, seagrass *T. testudinum* at 66±2%, and the green alga *H. opuntia* at 89±1%. Cholesterol was present in all primary producers except in mangrove leaves, and it was the main sterol in red algae at 30.2±0.1% and phytoplankton at 50±8%. Isofucosterol was the main sterol in brown algae at 37±1%, and it was shared with green algae at much lower proportions at 1.6±0.3%, while campesterol was the main sterol in zooxanthellae at 47±11%, and it was shared with seagrass, green and red algae, and phytoplankton (Table 3).

Table 3. Sterol composition of primary producers collected in the coral reef food web at PNSAV. Values are means ± 95% confidence interval. (H)= higher than all others, (S)= specific sterols, C= carbons and Δ= position of double bonds. Structure of sterols in bold are represented in Error! Reference source not found.. Superscript letters denote significant differences among columns. TS= Concentration of total sterols per dry weight.

df= degree of freedom, F= Fisher statistics, and p= Tukey's family error p-value

Sterol	Short formula	R. mangle	T. testudinum	H. opuntia	Galaxaura sp.	Dictyota sp.	Phytoplankton	Zooxanthellae	df	F	p-value
Pregnanone (S)	$C_{27}\Delta^0$	-	-	-	-	-	-	12±6	-	-	-
24-Nordehydrocholesterol (H)	$C_{26}\Delta^{5,22}$	3±1^a	0.7±0.2^{ab}	-	-	-	0.06±0.01^b	-	2, 14	35.0	<0.001
Trans-22-dehydrocholesterol (H)	$C_{27}\Delta^{22}$	-	5±1^a	-	-	-	0.60±0.01^b	-	1, 11	16.0	0.002
Occelasterol	$C_{27}\Delta^{5,22}$	-	-	-	-	-	7±3^a	6±3^a	1, 10	0.3	0.594
Trans-22-dehydrocholestanol (S)	$C_{27}\Delta^{22}$	-	-	-	-	-	0.6±0.5	-	-	-	-
Patinosterol (S)	$C_{27}\Delta^{22}$	-	-	-	-	-	0.4±0.5	-	-	-	-
Cholesterol (H)	$C_{27}\Delta^5$	-	3±1^a	0.9±0.1^a	30.2±0.1^b	1.3±0.4^a	50±8^c	29±4^b	5, 23	45.0	<0.001
Cholestanol (H)	$C_{27}\Delta^0$	-	-	-	11±2^a	-	5±1^b	-	1, 11	53.0	<0.001
Brassicasterol (H)	$C_{28}\Delta^{5,22}$	-	-	-	23±2^a	-	5±1^b	-	1, 11	30.7	<0.001
Brassicastanol (H)	$C_{28}\Delta^{22}$	-	-	-	5±1^a	-	0.9±0.7^b	-	1, 11	33.2	<0.001
Ergost-8(14)-enol (S)	$C_{28}\Delta^{24(28)}$	-	-	-	-	-	0.2±0.3	-	-	-	-
Stellasterol (H)	$C_{28}\Delta^{7,22}$	-	-	1.2±0.2^a	-	-	0.7±0.6^a	-	1, 13	1.04	0.328
24-Methylenecholesterol	$C_{28}\Delta^{5,22}$	-	-	0.9±0.2^a	-	-	6±1^b	-	1, 12	27.7	<0.001

Sterol	Short formula	R. mangle (H)	T. testudinum	H. opuntia	Galaxaura sp.	Dictyota sp.	Phytoplankton	Zooxanthellae	df	F	p-value
Campesterol (H)	$C_{28}\Delta^5$	-	9.4±0.4[ab]	4±2[a]	14±1[b]	-	9±2[ab]	47±11[c]	4, 20	77.5	<0.001
Stigmasterol (H)	$C_{29}\Delta^{5,22}$	8.1±2.7[a]	16±1[b]	1.2±1.0[c]	-	-	3±2[c]	-	3, 18	44.2	<0.001
Stigmastanol	$C_{29}\Delta^{22}$	-	-	-	-	-	0.7±0.8	-	-	-	-
Episterol (S)	$C_{28}\Delta^{7,24(28)}$	-	-	-	-	-	2±1	-	-	-	-
4,24 dimethyl 5,7-dien-3-β-ol (S)	$C_{29}\Delta^{5,7}$	-	-	-	-	19±2	-	-	-	-	-
Poriferasterol (S)	$C_{29}\Delta^{5,22}$	-	-	-	-	18±1	-	-	-	-	-
Spinasterol (S)	$C_{29}\Delta^{7,22}$	-	-	1.1±0.3	-	-	-	-	-	-	-
β-Sitosterol (H)	$C_{29}\Delta^5$	82±1[a]	66±2[b]	89±1[a]	17±1[c]	-	8±7[c]	-	4, 21	166.8	<0.001
Fucosterol (S)	$C_{29}\Delta^{5,24}$	-	-	-	-	21±1	-	-	-	-	-
Sitostanol (S)	$C_{29}\Delta^0$	-	-	-	-	-	2±2	-	-	-	-
Isomer of fucosterol (S)	$C_{29}\Delta^{5,24}$	-	-	-	-	5±1	-	-	-	-	-
Dinosterol (S)	$C_{30}\Delta^{22}$	-	-	-	-	-	0.4±0.3	-	-	-	-
Dinostanol (S)	$C_{30}\Delta^0$	-	-	-	-	-	0.5±0.4	-	-	-	-
Gorgosterol (S)	$C_{30}\Delta^5$	-	-	-	-	-	-	6±2	-	-	-
Isofucosterol (H)	$C_{29}\Delta^{5,24}$	-	-	1.6±0.3[a]	-	37±1[b]	-	-	1, 7	281.3	<0.001
Cycloartenol (S)	$C_{30}\Delta^5$	7±2	-	-	-	-	-	-	-	-	-
TS (mg g^{-1} dry wt)		0.9±0.2[a]	0.4±0.3[a]	0.4±0.3[a]	0.2±0.1[a]	4±3[a]	1.0±0.3[a]	15±4[b]	6, 26	54.6	<0.001

The mangrove *R. mangle* had 7±2% cycloartenol of total sterols, which is a precursor of phytosterols. Therefore, it could not be used as a biomarker sterol, while phytoplankton had several distinctive sterols. The seagrass *T. testudinum* had six sterols, and *trans*-22-dehydrocholesterol was distinctive at 5±1% of total sterols, being shared only with phytoplankton at <1%. The green alga *H. opuntia* had eight sterols; its distinctive sterols were spinasterol (1.1±0.3%) and stellasterol at 1.2±0.2%. Stellasterol was also present in phytoplankton samples at a significantly lower proportion. Similarly, the red alga *Galaxaura* sp. had three sterols shared with phytoplankton with significantly higher proportions in red algae: cholestanol (11±2%), brassicasterol (23±2%) and brassicastanol (5±1%).

The brown alga *Dictyota* sp. was composed of seven sterols, among which four sterols were distinctive: 4,24 dimethyl, 5,7 dien-3β-ol (19±2%), poriferasterol (18±1%), fucosterol (21±1%), and an isomer of fucosterol (5±1%).

Phytoplankton had 19 different sterols, among which eight were distinctive: episterol (2±1%), sitostanol (2±2%), 23,24 dimethylcholest 5,7-dien-3β-ol (0.7±0.8%), *trans*-22-dehydrocholestanol (0.6±0.5%), patinosterol (0.4±0.5%), dinosterol (0.4±0.3%), dinostanol (0.5±0.4%) and ergost 8(14)-enol (0.2±0.3%); however, they were all present at low proportions and had high variability because they came from different phytoplankton groups. Zooxanthellae had five sterols, among which pregnanone (12±6%) and gorgosterol (6±2%) were distinctive (Table 3).

Biomarkers were selected from distinctive primary producer sterols that were detected in zooplankton, the sponge *Aplysina* sp., sea pen shell *P. carnea*, coral *M. cavernosa* and the sea urchin *E. lucunter*. Fish were not considered to define biomarkers because they can synthesize cholesterol from other sterols and absorb phytosterol to a lesser extent. For instance, pregnanone, ergost-8(14)-enol, spinasterol, and fucosterol were distinctive for zooxanthellae, phytoplankton, green and brown algae, respectively.

However, they cannot be considered as biomarkers because they were not incorporated in the zooplankton, sponge *Aplysina* sp., sea pen shell *P. carnea*, coral *M. cavernosa* or sea urchin *E. lucunter* tissues. In contrast, cycloartenol from mangrove was detected in zooplankton in the rainy season with a variable and low proportion of 0.4±0.7%, but it is a phytosterol precursor; therefore, it could not be a biomarker. On the other hand, the seagrass biomarker *trans*-22-dehydrocholesterol was detected in zooplankton.

Stellasterol from the green alga *H. opuntia* and phytoplankton was identified in zooplankton, *E. lucunter*, *P. carnea* and *Aplysina* sp. Sterols from the brown alga *Dictyota* sp., 4,24 dimethyl 5,7 dien-3β-ol and poriferasterol, were detected in *E. lucunter* and the sponge *Aplysina* sp. Isofucosterol was also detected in the sponge *Aplysina* sp. In turn, brassicasterol from the red alga *Galaxaura* sp. and phytoplankton was detected in all invertebrates studied, with the highest proportions in the sponge *Aplysina* sp. ($F_{9,\,39}$=29.6, p<0.001) at 20±1% (Table 4).

Occelasterol was detected in phytoplankton and zooxanthellae, while episterol was distinctive in phytoplankton and gorgosterol in zooxanthellae (Table 3).

Occelasterol was better incorporated in zooplankton than episterol, and it was probably consumed by the great star coral *M. cavernosa*. Occelasterol was also incorporated into the masked goby *C. personatus*. The distinctive sterol from phytoplankton, episterol, was detected in all invertebrates studied, except in the coral *M. cavernosa,* the latter contained gorgosterol, a distinctive sterol of zooxanthellae (Table 4).

The green algae biomarker, stellasterol was detected in all teleost fish except in the hogfish *B. rufus*. The phytoplankton biomarker episterol was detected in all fish with a higher proportion in the masked goby *C. personatus*.

The brown algae biomarkers, fucosterol and isofucosterol, were detected in *B. rufus,* a sea urchin predator, and fucosterol was also detected in *C. personatus* a zooplanktivore goby. Primary producer sterols such as β-sitosterol were detected in all fish studied (Table 5).

Table 4. Sterol composition of zooplankton, coral, sea urchins, clams and sponges collected during the dry season in PNSAV. Values are means ± 95% confidence intervals

Sterol	Short formula	Zooplankton	M. cavernosa	E. lucunter	P. carnea	Aplysina sp.	df	F or H	p-value
24-Nordehydrocholesterol	$C_{26}\Delta^{5,22}$	0.9±0.2[a]	-	-	3±1[b]	-	-	-	<0.05
Trans-22-dehydrocholesterol	$C_{27}\Delta^{22}$	6±2	-	-	-	-	-	-	-
24-Nor-22,23 methylenecholest-5-en-3-β-ol	$C_{27}\Delta^{5}$	-	-	-	2±1[ab]	-			0.056
Occelasterol	$C_{27}\Delta^{5,22}$	2±1[a]	2±1[a]	2±1[a]	9±2[c]	5±1[b]	9	24	0.001
Trans-22-dehydrocholestanol	$C_{27}\Delta^{22}$	2.5±0.4	-	-	-	-	-	-	-
Cholesterol	$C_{27}\Delta^{5}$	51±5[a]	14± 4[b]	85±2[d]	38±10[a]	25±1[c]	9, 39	28.1	<0.001
Cholestanol	$C_{27}\Delta^{0}$	3± 1[ab]	-	-	2±1[b]	-			<0.050
Desmosterol	$C_{27}\Delta^{5,24}$	-	-	-	-	0.6±0.1[a]			>0.050
7-Dehydrocholesterol	$C_{28}\Delta^{5,7}$	-	-	0.7±0.3[a]	-	-			>0.050
Secocholesta-5(10), 6,8 triene	$C_{27}\Delta^{5,6,8}$	-	-	-	2 ±1[a]	-			>0.050
Brassicasterol	$C_{28}\Delta^{5,22}$	6±2[a]	1±1[b]	3±1[b]	14±2[c]	20±1[d]	9, 39	29.6	0.001
Brassicastanol	$C_{28}\Delta^{22}$	2±1[a]	-	-	0.9±0.5[a]	-			>0.05
Ergost-7-enol	$C_{28}\Delta^{7}$	-	-	-	-	-			>0.05
Ergosterol	$C_{28}\Delta^{5,7,22}$	-	-	-	1.3±0.3[a]	-			<0.05
Stellasterol	$C_{28}\Delta^{7,22}$	-	-	0.4±0.2[a]	2±1[b]	1.9±0.4[b]	8	17	0.009
24- Methylenecholesterol	$C_{28}\Delta^{5,24}$	3±1[a]	-	-	3±3[a]	-	3, 22	2.8	0.061
Dihydrobrassicasterol	$C_{28}\Delta^{5}$	-	-	0.6±0.1[a]	-	-			>0.05
Campesterol	$C_{28}\Delta^{5}$	4±1[a]	53±5[c]	2.1±0.2[a]	6±2[ab]	4.6±0.4[ab]	9, 37	116	<0.001
Stigmasterol	$C_{29}\Delta^{5}$	0.5±0.2[a]	0.3±0.2[a]	2±1[a]	2±1[ab]	6.5±0.2[c]	9, 40	36.5	<0.001
Stigmastanol	$C_{29}\Delta^{22}$	2.1±0.4	-	-	-	-	-	-	-
Episterol	$C_{28}\Delta^{7,24(28)}$	-	0.2±0.1[b]	0.6±0.3[a]	2.4±0.4[a]	2.3±0.4[a]	8, 30	2.7	0.025

Sterol	Short formula	Zooplankton	M. cavernosa	E. lucunter	P. carnea	Aplysina sp.	df	F or H	p-value
4-24 dimethyl 5,7-dien-3-β-ol	$C_{29}\Delta^{5,7}$	-	-	0.5± 0.1^a	-	4.9±0.1^b			<0.050
Poriferasterol	$C_{29}\Delta^{5,22}$	-	-	0.4±0.1^a	-	4±1^b	3, 11	47	0.001
β-Sitosterol	$C_{29}\Delta^{5}$	6±2^{ab}	1.1±0.8^a	3.1±0.2^{ab}	10±6^{ab}	15.4±0.3^b	9, 39	12.5	0.001
Sitostanol	$C_{29}\Delta^{0}$	1.2± 0.4^a	2±1^a	-	-	-			>0.05
4, 24 dimethyl 5α cholestan 3-β-ol	$C_{29}\Delta^{0}$	-	-	-	0.3± 0.1^a	-			>0.05
Isomer of fucosterol	$C_{29}\Delta^{5,24}$	-	-	-	-	1.6±0.3^a			>0.05
4,23,24 trimethyl 5α cholest 24(28)-en-3β-ol	$C_{30}\Delta^{24(28)}$	-	-	-	3± 2^a	-			>0.05
Dinosterol	$C_{30}\Delta^{22}$	1.0±0.4^a	0.3±0.2^b	-	-	-			<0.050
Dinostanol	$C_{30}\Delta^{0}$	-	7±6^a	-	-	-			>0.050
4,23,24 trimethyl-5α -cholest-7-en-3β-ol	$C_{30}\Delta^{7}$	-	-	-	0.4±0.1^a	-			<0.050
Gorgosterol	$C_{30}\Delta^{5}$	-	20±4^a	-	-	-	1, 6	11.2	0.016
Isofucosterol	$C_{29}\Delta^{5,24}$	-	-	0.7±0.2^a	-	-			>0.050
Isomer of fucosterol	$C_{29}\Delta^{5,24}$					1.6±0.3^a	1, 5	0.346	0.582
Cycloartenol	$C_{30}\Delta^{24}$	-	-	-	-	-			
TS (mg g^{-1} dry wt)		3±1^a	13±2^b	3±3^a	12±3^b	10±4^b	9, 39	13.3	<0.001

Table 5. Sterol composition of fish collected in the PNSAV. Values are means ± 95% confidence intervals. Superscript letters denote significant differences among columns. Boldface compounds originate in zooplankton (Zoo) and are present only in one fish taxon (Tax)

Sterol	Short formula	A. chirurgus	C. personatus	B. rufus	O. chrysurus	C. hippos	F	p value
24-Nordehydrocholesterol	$C_{26}\Delta^{5,7}$	-	0.5±0.1[a]	0.8±0.7[a]	0.3±0.2[a]	0.6±0.3[a]	CI	<0.05
Trans 22 dehydrocholesterol	$C_{27}\Delta^{22}$	-	3±2[a]	-	-	-		
Occelasterol	$C_{27}\Delta^{5,22}$	-	2.9±0.2[a]	0.2±0.1[c]	0.3±0.2[c]	0.10±0.03[c]	24	0.001
Cholesterol	$C_{27}\Delta^{5}$	92±1[a]	83±2[b]	94±1[a]	96.1±1.1[d]	96.3±0.1[d]	28.1	0.001
Cholestanol	$C_{27}\Delta^{0}$	-	1.9±0.1[a]	0.9±0.2[b]	1.0±0.2[a]	0.6±0.1[b]		0.040
Brassicasterol	$C_{28}\Delta^{5,22}$	-	0.42±0.03[a]	0.4±0.3[a]	0.2±0.1[a]	0.07±0.04[b]	2.8	0.020
Ergost-7-enol (Zoo)	$C_{28}\Delta^{7}$	0.6±0.3[a]	-	0.5±0.2[a]	0.6±0.3	0.3±0.1[a]		
Stellasterol	$C_{28}\Delta^{7,22}$	0.9±0.3[a]	0.8±0.2[a]	-	0.5±0.1[a]	0.2±0.1[a]	17	0.009
24- Methylenecholesterol	$C_{28}\Delta^{5,24}$	-	1.8±0.1[a]	0.4±0.1[a]	0.5±0.1[a]	0.3±0.1[a]	12.7	0.001
Campesterol	$C_{28}\Delta^{5}$	1.2±0.5[a]	2.1±0.1[a]	0.8±0.4[a]	0.31±0.04[c]	0.5±0.1[c]	14.5	0.001
5α-Cholest-7-en-3β-ol (Tax) (Lathosterol)	$C_{27}\Delta^{7}$	-	0.3±0.1[a]	-	-	-		0.050
Stigmasterol	$C_{29}\Delta^{5}$	-	0.34±0.01[a]	0.2±0.1[a]	0.2±0.2[a]	0.6±0.1[c]	CI	<0.050
Episterol	$C_{28}\Delta^{7,24(28)}$	0.9±0.5[a]	1.8±0.2[b]	0.3±0.1[a]	0.3±0.1[a]	0.14±0.05[a]	CI	<0.050
β-Sitosterol	$C_{29}\Delta^{5}$	5±2[a]	1.3±0.1[b]	0.4±0.1[b]	0.04±0.02[b]	0.20±0.04[b]	CI	<0.050
Fucosterol	$C_{29}\Delta^{5,24}$	-	0.40±0.02[a]	0.08±0.03[b]	-	-		
Sitostanol	$C_{29}\Delta^{0}$	-	-	-	-	-		
Dinostanol	$C_{30}\Delta^{22}$	-	-	0.08±0.03[a]	-	-		
Isofucosterol	$C_{29}\Delta^{5,24}$	-	-	0.04±0.02[a]	-	-		
TS (mg g^{-1} dry wt)		2±1[a]	3±2[a]	7±1[b]	3±2[a]	2±1[a]	CI	<0.050

TS= Concentration of total sterols per dry weight. df= degree of freedom, F= Fisher statistics or H= Kruskal-Wallis statistic, and p= Tukey's family error p value. CI= Confidence interval was used to determine significant differences.

3.5.2. Accumulation and Correlation of Structural and Storage Lipid Classes with Sterol Concentrations

The concentration of structural lipids like phospholipids and free sterols in consumer tissues did not change seasonally. The only exception was the hogfish *B. rufus* with the highest concentration of sterols in the dry season ($F_{9, 26}=7.0$, $p<0.001$) at 5 ± 1 mg g^{-1} (Carreón-Palau et al. 2018). The increase of sterols in the hogfish *B. rufus* is explained by the concentration of cholesterol and of fucosterol to a lesser extent. On the other hand, the phytosterol *β*-sitosterol had the highest concentration in the mangrove *R. mangle* at 0.8 ± 0.1 mg g^{-1}.

In second place was the seagrass and green algae with *β*-sitosterol concentrations of 0.4 ± 0.2 mg g^{-1}, the lowest concentration was detected in phytoplankton at 0.1 ± 0.1 mg g^{-1} ($F_{6, 26}=13.3$, $p<0.001$) and finally zooxanthellae had none detected. Suspension feeders such as bivalves *P. carnea* and sponges *Aplysina* sp. showed an accumulation of *β*-sitosterol increasing their concentration between 10 and 30 times with respect to phytoplankton and between two to four times with respect to mangrove. This accumulation was significantly higher in the sponge. In contrast, fish had a trophic reduction in *β*-sitosterol with low concentrations ranging between 0.003 and 0.1 mg g^{-1}.

Apportionment of brown algae was recorded using poriferasterol, a biomarker observed in the sea urchin and sponge, while fucosterol was detected in the masked goby and hogfish. Among these, sea urchins and fish had lower concentrations than brown algae, while sponges had similar concentrations (Table 6), suggesting an accumulation by filter feeding.

In the present study, the whole organism was analyzed including spat tissues of *P. carnea*, *Aplysina* sp. and *M. cavernosa* and they increased their concentration of episterol compared to phytoplankton ($F_{9, 39}=9.9$, $p<0.001$). Meanwhile, fish had a similar concentration of 24-methylenecholesterol compared to phytoplankton. Similarly, gorgosterol the zooxanthellae biomarker was detected only in the great star coral *M. cavernosa* with significantly higher concentrations compared to the source zooxanthellae (Table 6).

Table 6. Concentration of distinctive sterols from primary producers (mg g^{-1} dry wt). Values are mean ± standard deviation. F= Fisher statistic, p= probability value

Primary producers/ biomarkers	β-sitosterol	Poriferasterol and fucosterol	Episterol and 24-methylenecholesterol	Gorgosterol	Cholesterol
R. mangle	0.8±0.1^c				
T. testudinum	0.3±0.2^{ab}				0.02±0.01^a
H. opuntia	0.4±0.2^b				0.02±0.01^a
Galaxaura sp.	0.1±0.1^a				0.10±0.06^a
Dictyota sp.		0.9±0.5			
Phytoplankton	0.1±0.1^a		0.02±0.01		0.4±0.3^b
Symbionts					
Zooxanthellae				0.8±0.3	4±1^c
$F_{6, 26}$	13.27	16.52	3.79	42.88	146.11
p	< 0.001	<0.001	0.008	<0.001	<0.001
Invertebrates	β-sitosterol	Poriferasterol	Episterol	Gorgosterol	Cholesterol
Herbivores					
E. lucunter	0.1±0.1^a	0.01±0.01^a	0.01±0.01^a		3±2^a
P. carnea	1.1±0.5^{ab}		0.3±0.1^b		4±2^{bc}
Aplysina sp.	1.7±0.6^b	0.6±0.1^b	0.3±0.01^b		3±1^b
Planktivores					
Zooplankton	0.3±0.2^a				1.7±1.6^a
Symbionts					
M. cavernosa	0.2±0.1^a		0.02±0.01^a	2.3±0.4^a	1.6±1.2^a
$F_{4, 24}$	38.79	72.32	9.92		7.02
P	< 0.001	< 0.001	<0.001		<0.001
Fish	β-sitosterol	Fucosterol	24-Methylenecholesterol	Gorgosterol	Cholesterol
Herbivores/ detritivores					
A. chirurgus	0.100^a		0.02±0.01^a		1.6±0.7^a
Planktivores					
C. personatus	0.050^b	0.02±0.002^a	0.07±0.01^b		3.4±0.5^{ab}
Mollusc-eating/ Echinoderm-eating					
B. rufus	0.030^b	0.01±0.002^a	0.03±0.01^a		6.8±1.0^c
Piscivores					
O. chrysurus	0.0013^b		0.02±0.01^a		2.9±1.6^{ab}
C. hippos	0.003^b		<0.005^a		1.6±1.2^a
$F_{4, 11}$	7.89		4.92	-	10.69
p	< 0.001		0.001	-	<0.001

Table 7. Pearson's correlation with p-values (in brackets) among concentration (mg g^{-1} dry wt.) of biomarker and *de novo* sterols and storage and structural lipid classes. Asterisks denote significant correlations

Sterol	Source	Invertebrates (Zooplankton, clam, sea urchin, coral, and sponge)						Teleost fish (masked goby, surgeonfish, hogfish, yellowtail snapper and jack)					
		WE+SE	TAG	ST	AMPL	PL	TAG:ST	Result	WE+SE	TAG	ST	AMPL	PL	TAG:ST	Result
Cycloartenol	*R. mangle*	-0.080 (0.585)	-0.041 (0.780)	-0.095 (0.515)	-0.033 (0.822)	-0.056 (0.704)	-0.017 (0.908)	no	-	-	-	-	-	-	-
Trans-22-dehydrocholesterol	*T. testudinum*	-0.118 (0.441)	0.091 (0.532)	0.123 (0.399)	-0.261 (0.070)	-0.073 (0.620)	-0.055 (0.706)	no	0.225 (0.187)	-0.166 (0.333)	0.459* (0.005)	0.271 (0.110)	0.598* (0.001)	-0.197 (0.250)	no
Stellasterol	*H. opuntia*	-0.181 (0.213)	-0.153 (0.295)	0.687* (0.001)	0.373* (0.008)	0.473* (0.001)	-0.266 (0.064)	no	0.148 (0.390)	-0.286 (0.090)	0.506* (0.002)	0.132 (0.443)	0.215 (0.208)	-0.392* (0.018)	no
Brassicasterol	*Galaxaura* and phytoplankton	-0.081 (0.579)	-0.209 (0.150)	0.742* (0.001)	0.352* (0.013)	0.488* (0.001)	-0.350* (0.014)	no	0.407* (0.014)	0.052 (0.761)	0.744* (0.001)	0.498* (0.002)	0.384* (0.021)	-0.066 (0.702)	yes
Poriferasterol	*Dictyota* sp.	-0.120 (0.412)	-0.128 (0.382)	0.794* (0.001)	0.458* (0.001)	0.516* (0.001)	-0.234 (0.105)	no	--	--	-	-	-	-	-
Fucosterol	*Dictyota* sp.	-0.115 (0.431)	-0.156 (0.286)	0.795* (0.001)	0.466* (0.001)	0.510* (0.001)	-0.257 (0.074)	no	0.225 (0.188)	-0.122 (0.480)	0.406* (0.014)	-0.169 (0.326)	-0.080 (0.641)	-0.119 (0.489)	no
4-24 dimethyl 5,7 dien-3-β-ol	*Dictyota* sp.	-0.123 (0.401)	-0.115 (0.431)	0.795* (0.001)	0.460* (0.001)	0.512* (0.001)	-0.220 (0.128)	no	-	-	-	-	-	-	-
Episterol	phytoplankton	-0.070 (0.597)	-0.061 (0.678)	0.555* (0.001)	0.309* (0.031)	0.314 (0.208)	-0.195 (0.179)	no	-0.005 (0.715)	-0.292 (0.085)	0.153 (0.373)	-0.158 (0.357)	0.166 (0.332)	-0.333* (0.047)	no
Gorgosterol	zooxanthellae	0.865* (0.001)	-0.086 (0.569)	-0.200 (0.160)	0.099 (0.499)	-0.217 (0.134)	-0.010 (0.945)	yes	-	-	-	-	-	-	-

Table 7. (Continued)

Sterol	Source	Invertebrates (Zooplankton, clam, sea urchin, coral, and sponge)						Teleost fish (masked goby, surgeonfish, hogfish, yellowtail snapper and jack)					
		WE+SE	TAG	ST	AMPL	PL	TAG:ST	Result	WE+SE	TAG	ST	AMPL	PL	TAG:ST	Result
β-sitosterol	R. mangle, T. testudinum and H. opuntia	-0.081 (0.586)	-0.155 (0.288)	0.782* (0.001)	0.345* (0.015)	0.462* (0.001)	-0.307* (0.032)	no	0.021 (0.487)	0.141 (0.412)	0.215 (0.207)	0.262 (0.123)	0.167 (0.332)	-0.067 (0.696)	no
Occelasterol	Phytoplankton and zooxanthellae	0.186 (0.200)	-0.217 (0.135)	0.311* (0.029)	0.354* (0.012)	0.238 (0.100)	0.263 (0.068)	no	0.205 (0.230)	-0.124 (0.472)	0.644* (0.001)	0.683* (0.001)	0.543* (0.001)	-0.177 (0.301)	no
24-methylenecholesterol	Phytoplankton and H. opuntia	-0.200 (0.169)	-0.055 (0.705)	-0.023 (0.877)	-0.097 (0.509)	-0.025 (0.862)	-0.162 (0.266)	no	-0.090 (0.603)	0.518* (0.001)	0.182 (0.287)	0.191 (0.264)	0.177 (0.302)	0.263 (0.121)	yes
Cholesterol	De novo, Galaxaura sp.	0.011 (0.650)	0.047 (0.748)	0.620* (0.001)	0.156 (0.286)	0.517* (0.001)	-0.254 (0.074)	no	0.280 (0.098)	-0.283 (0.094)	0.816* (0.001)	0.150 (0.384)	0.317 (0.059)	-0.427* (0.009)	no

Pearson correlation tests were performed to verify the coincidence of primary producer sterol biomarkers with the highest storage lipid classes such as TAG and WE+SE, as well as the TAG:ST ratio (Table 7). Macroalgae and phytoplankton sterols were significantly positively correlated with structural lipids such as ST, AMPL and PL, but not with storage lipids for invertebrates such as zooplankton, clam *P. carnea*, sea urchins *E. lucunter* and sponge *Aplysina* sp. The only sterol with a significant positive correlation ($r=$ 0.865, $p<$ 0.001) between concentrations was gorgosterol and wax and steryl esters present in the coral *M. cavernosa* (Table 7). In contrast, fish had a significant positive correlation between brassicasterol from red algae and phytoplankton with storage wax and steryl esters, as well as with structural lipids. Meanwhile, 24-methylenecholesterol from phytoplankton was the only biomarker significantly positively correlated ($r=0.518$, $p=0.001$) with TAG confirming that phytoplankton had the highest nutritional quality for the PNSAV fish.

3.6. Discussion

3.6.1. Accumulation and Correlation of Structural and Storage Lipid Classes with Primary Producer Source Sterols in the coral Reef Food Web

Sterol biosynthesis in invertebrates does not take place or else proceeds at a slow rate. For instance, the sterol components of sponges are likely to vary with habitat. While bivalves may be capable of sterol biosynthesis (Kanazawa, 2001 and references therein). For instance, the incorporation of unmodified dietary phytosterols was detected after a 6-week feeding period reflecting the diet composition, principally in spat tissues of bivalves (Kanazawa, 2001). In addition, Basen et al. (2012) provided evidence that somatic growth of the bivalve *C. fluminea* on cyanobacterial diets is constrained by the absence of sterols, as indicated by a growth-enhancing effect of sterol supplementation, suggesting that sterols are potentially limiting nutrients in aquatic food webs.

The concentration of sterols in the whole organism of bivalve *P. carnea* and coral *M. cavernosa* suggests differential sterol retention depending on feeding habits: filter-feeding *versus* symbiotic apportionment. While *M. cavernosa* depends on zooxanthellae contributions, *P. carnea* reflects the available phytoplankton. On the other hand, fish can maintain constant concentrations of cholesterol because they synthesize cholesterol from acetyl-Co A and mevalonic acid (Leaver et al., 2008). According to our results, phytoplankton was the primary producer with the highest nutritional quality. This source promoted a better condition in teleost fish, bivalves, and zooplankton. For coral, gorgosterol-traced zooxanthellae promoted a better nutritional condition.

CONCLUSION

Coral reef sterols were identified by GC-MS allowing distinction among peaks running at similar retention times for different samples. Once peaks were identified by MS, the same peaks were detected with GC-FID and quantified by GC-FID. These instruments revealed 29 sterols occur in primary producers and 37 sterols occur in their consumers. Sterols from primary producers allowed the identification of the main sources of organic carbon in the coral reef food web.

Red and green algae sterols were shared with phytoplankton making their identification difficult. In contrast, brown algae distinctive sterols were found in sea urchins and sponges. Cholesterol showed significantly higher concentrations in all fish. In contrast to teleost fish, the great star coral *M. cavernosa* showed a better nutritional condition as reflected by higher wax and steryl ester proportions that correlated well with gorgosterol-traced zooxanthellae. The phytoplankton sterol episterol was correlated with higher contents of storage lipid TAG. In addition, an increase in the proportion of *trans*-22-dehydrocholesterol, the characteristic sterol of seagrass, and trace amounts of 24-nordehydrocholesterol from mangrove were detected.

REFERENCES

Ackman, G. R. (2002). The gas chromatograph in practical analyses of common and uncommon fatty acids for the 21st century. *Analytica Chimica.* 465:175-192.

Agilent Technologies (2012). Agilent GC/MSD ChemStation and instrument operation. Course number I (H4043A). *Agilent Technologies,* INC., USA. pp.3-10.

Basen, T., Rothhaupt, K. O., Martin-Creuzburg, D. (2012). "Absence of sterols constrains food quality of cyanobacteria for an invasive freshwater bivalve." *Oecologia,* 170:57–64.

Carreón-Palau, L., Parrish, C. C., Del Angel-Rodríguez, J. A., Pérez-España, H. and Aguíñiga-García, S. (2013). "Revealing organic carbon sources fueling a coral reef food web in the Gulf of Mexico using stable isotopes and fatty acids." *Limnology and Oceanography,* 58(2): 593–612.

Carreón-Palau, L. (2015). Organic carbon sources and their transfer in a Gulf of Mexico coral reef ecosystem. Doctoral (Ph.D.) thesis, Memorial University of Newfoundland. 268 pp. http://research.library.mun.ca.

Carreón-Palau, L., Parrish, C. C., Pérez-España, H., and Aguíñiga-García, S. (2018). "Elemental ratios and lipid classes in a coral reef food web under river influence." *Progress in Oceanography,* 164:1-11.

Christian, G. D. (1981). *Química analítica.* 2a Edición. Limusa México. [Analytical Chemistry. Second edition. Limusa Mexico] 684 pp.

Christhie, W. W. (2003). *Lipid Analysis: Isolation, Separation, Identification and Structural Analysis of Lipids.* Bridgater, England. 416 pp.

Christhie, W. W. and Xianlin, H. (2012). *Lipid Analysis: Isolation, Separation, Identification and lipidomic analysis.* 4th edition Woodhead Publishing Limited. Middlesex, UK. 409 pp.

Cohen, Z., Norman, H. A., and Heimer, Y. M. (1995). "Microalgae as a source of omega-3 fatty acids." In: Simopoulos, A. P. (editor) *World Rev Nutr Diet, Plants in Human Nutrition,* Karger, Basel. 77: 1-31.

Crandall, J. B., Teece, M. A., Estes, B. A., Manfrino, C., and Ciesla, J. H. (2016). "Nutrient acquisition strategies in mesophotic hard corals using compound specific stable isotope analysis of sterols." *J. Exp. Mar. Biol. Ecol.* 474: 133–141.

García-Llatas, G. and Rodríguez-Estrada, M. T. 2011. "Current and new insights on phytosterol oxides in plant sterol-enriched food." *Chemistry and physics of lipids,* 164:607-624.

Goad L. J. (1981). "Sterol biosynthesis and metabolism in marine invertebrates." *Pure App. Chem.* 51: 837-852.

Gunstone, F. D., (1996). *Fatty Acid and Lipid Chemistry.* London: Blackie Academic & Professional. 252 pp.

Jones, G. J., Nichols, P. D., and Shaw, P. M., (1994). "Analysis of microbial sterols and hopanoids." In: Goodfellow, M., O'Donnell, A. G. (Eds.), *Chemical methods in prokaryotic systematics.* John Wiley, Chichester, pp. 163–195.

Kanazawa A., 2001. "Sterols in marine invertebrates." *Fisheries Science,* 67: 997–1007.

Kitson, F. G., Larsen, B. S. & McEwen, C. N. (1996). *Gas chromatography and mass spectrometry.* Academic Press, USA. 9-8 pp.

Leaver, M. J., Villeneuve, L. A. N., Obach, A., Jensen, L. Bron, J. E. Tocher, D. R., and Taggart, J. B. 2008. "Functional genomics reveals increases in cholesterol biosynthetic genes and highly unsaturated fatty acid biosynthesis after dietary substitution of fish oil with vegetable oils in Atlantic salmon (Salmo salar)." *BMC Genomics,* 9:299, doi: 10.1186/1471-2164-9-299.

Lederer, E. & M. Lederer, 1957. *Chromatography, a review of principles and applications.* 2nd Ed. Elsevier, Amsterdam. 711pp.

Martin-Creuzburg, D., and von Elert, E. 2009. "Ecological significance of sterols in aquatic food webs." In: Kainz, M., Brett, M., and Arts, M. (eds.), *Lipids in Aquatic Ecosystems,* Springer New York, NY. doi: 10.1007/978-0-387-89366-2_3.

Martin-Creuzburg D., Oexle, S. and Wacker, A. J. 2014. "Thresholds for sterol-limited growth of Daphnia magna: A comparative approach using 10 different sterols." *J. Chem. Ecol.* 40:1039-1050.

McLafferty, F. W., and Tureček, F. (1993). *Interpretation of mass spectra*. Fourth edition. University Science Books, Mill Valley, California. 23(6): 379 pp.

McNair, H. M. 1981. *Cromatografía de gases. Secretaría General de la OEA*. Monografía No. 23. Serie Química. [Gas Chromatography. Organization of American States. Monograph 23. Chemistry Series]. Washington D.C. 90 pp.

McNair, H. M., and Miller J. M. (1997). *Basic gas chromatography*. New York, E.U. A. 101-123.

Moreau, R. A., Whitaker, B. D., Hicks, K. B. 2002. "Phytosterols, phytostanols, and their conjugates in foods: structural diversity, quantitative analysis, and health-promoting uses." *Prog. Lipid Res.* 41:457–500.

Parrish, C. C., T. A. Abrajano, S. M. Budge, R. J. Helleur, E. D. Hudson, K. Pulchan Y C. Ramos. 2000. "Lipid and phenolic biomarkers in marine ecosystems: analysis and applications." In: P. Wangersky (Ed.) *The handbook of environmental chemistry*. Vol. 5, Part D. Springer-Verlag, Berling. 193-223 Pp.

Ramprasath, V. R. and Awad, A. B. (2015). "Role of phytosterols in cancer prevention and treatment." *Journal of the AOAC International*. 98 (3): 735–738.

Salas-Monreal D., Salas-de-León, D. A., Monreal-Gómez, M. A. and Riverón-Enzástiga, M. L. 2009. "Current rectification in a tropical coral reef system." *Coral Reefs*. 28:871–879. doi:10.1007/s00338-009-0521-9.

Siuzdak, G. (1996). Mass spectrometry for biotechnology. San Diego, Calif, Academic Press. 161 pp.

In: Sterols: Types, Classification and Structure ISBN: 978-1-53617-231-7
Editor: Scott Jimenez © 2020 Nova Science Publishers, Inc.

Chapter 2

FRESHWATER SPONGE STEROLS

Iuri Bezerra de Barros[1],
Glaucia Cristina Manço da Costa Bolson[1]
and Valdir Florencio da Veiga Junior[1,2]

[1]Chemistry Department, Amazonas Federal University,
Manaus, Amazonas, Brazil
[2]Chemical Engineering Department,
Military Institute of Engineering, Rio de Janeiro,
Rio de Janeiro, Brazil

ABSTRACT

Sponges are ancient and simple animals with metabolic richness that has caught the attention of the scientific community. With more than 8,000 species, sponges are abundantly observed in marine, as well as freshwater, environments. Studies regarding the chemical composition and pharmacological properties studies of these organisms mainly focus on the marine species. There are four phyla, with only species of one, the Demospongiae, found in freshwater environments. The species within Demospongiae can be further divided into six living families and only fossil records for a seventh. Among the metabolites of the sponge are the sterols. These are of particular importance due to their great diversity, and thus huge biotechnological potential. Sterols have been described from different sponge families from different continents, however, few studies have been conducted involving the sterols of the freshwater sponge compared to those of the marine environment. The first report of sponge sterols dates back to 1941, which described the presence of 5,6-dihydrostigmasterol in the species *Spongilla lacustris*. Since then, it has been shown that sponges can acquire their sterols through different processes, such as absorption and modification, which highlights the influence of the medium on the diversity of these metabolites. The number of sterols present varies with the species, not being uncommon the presence of a major sterol with the others appearing in trace concentration. These characteristics make it possible to use this data as chemosystematic markers.

Keywords: Freshwater sponges, Amazonia, Baikal lake, Demospongiae

INTRODUCTION

Sponges are the most primitive multicellular animals, with a fairly simple organization. Although they present specialized cells, these are not organized into organs or tissues (Ruppert, Fox and Barnes 1996; Hickman Jr., Roberts and Larson 2000). Exclusively aquatic and benthic, sponges filter microscopic-size food particles from dissolved organic matter in the water (de Goeji et al. 2008).

Encompassing more than 8,000 species, the Porifera phylum is divided into four classes, Demospongiae, Hexactinelidae, Homoscleromorpha and Calcarea. This biodiversity reflects its simple cellular organization associated with the great tolerance to symbiont microorganisms in sponges. These characteristics enable many evolutionary solutions, allowing the colonization of different environments (Van Soest et al. 2012; 2019).

The Demospongiae is the class with the greatest diversity and the only one whose species live in the freshwater environment. This class has more than 200 living species, which can be divided into six families: Lubomirskiidae, Malawispongiidae, Metaniidae, Metschnikowiidae, Potamolepidae and Spongillidae. A seventh family, Alaeospongillidae, only contains fossilized records (Manconi and Prozato 2008).

Like the well-studied marine sponges, which have been reported to possess several bioactive compounds, such as alkaloids, macrolides, peptides, polykettes, quinines, sterols and terpenoids (Mehbub et al. 2014), species of freshwater sponges have been described as chemically constituted by bioactive substances, such as fatty acids, and taxonomic biomarkers, such as sterols (de Barros, Volkmer-Ribeiro and da Veiga Junior 2013).

Several publications have elucidated the chemical composition of freshwater sponge species, indicating that they are rich in silica, as well as presenting a variety of other minerals, such as calcium, iron, copper, zinc, sulfur, aluminum, chlorine, titanium, vanadium, manganese, strontium, zirconium, potassium and sodium. The presence of these minerals proves that freshwater sponges species have a flexible capacity for absorption of substances from the environment (de Barros, Volkmer-Ribeiro and da Veiga Junior 2013).

The majority of the freshwater sponge publications explored the fatty acid and sterol composition of sponges from Lake Baikal, Russia. Furthermore, a wide variety of lipids with unusual features have been observed in freshwater sponges (de Barros, Volkmer-Ribeiro and da Veiga Junior 2013).

The secondary metabolism of invertebrates and freshwater microorganisms has been investigated for the isolation of biologically active chemicals, and also for novel structures or relevant taxonomies. Due to the richness and diversity of their metabolites, sponges are an excellent source of natural products (de Barros, Volkmer-Ribeiro and da Veiga Junior 2013).

SPONGES NATURAL PRODUCTS

The metabolic diversity of sponges is remarkable, and hundreds of new substances are isolated from these organisms annually.

In five years alone over a 1000 were isolated with 231 in 2017 (Carrol et al. 2019), 224 in 2016 (Blunt et al. 2018), 291 in 2015 (Blunt et al. 2017), 283 in 2014 (Blunt et al. 2016), and 243 in 2013 (Blunt et al. 2015). Since natural products have great relevance in the area of drug discovery for all diseases and illnesses, the large concentration of substances obtained from sponges make them a hot spot for this search (Jiménez 2018; Pereira 2019).

Sponges acquire their sterols by four different mechanisms: direct biosynthesis; ingestion; modification of ingested sterols; or biosynthesis associated with a symbiotic organism (Goad 1983; Djerassi and Silva 1991). The metabolic arsenal of marine and freshwater sponges able to convert 24-alkyl substituted into cholesterol, and surprisingly the reverse process also shows (Malik, Kerr and Djerassi 1988; Kerr et al. 1990; Kerr et al. 1992).

With just a basic backbone there are a number of unusual structures of the sponge sterols with odd branching patterns, which gives rise to the different biological activities (Dembitsky, Rezanka and Srebnik 2003; Gallimore et al. 2008). Contignasterol, for example, was first isolated from *Petrosia contignata*. It is a highly oxygenated sterol which is interesting due to the unusual proton 14β configuration and hemiacetal functionality in the side chain. This sterol, and some of its derivatives, exhibit anti-inflammatory activity, and is in preclinical tests for its antiasthmatic effects (Burgoyne, Andersen and Allen 1992; Gross and König 2006; Cheung et al. 2016).

FRESHWATER SPONGES' NATURAL PRODUCTS

The research on the natural products of freshwater sponges is more limited than for marine sponges. These studies have investigated the lipids present in these organisms, as well as uncommon natural products. Freshwater sponges rarely present alkaloids, terpenoids or peptides. The primary chemical organic composition lies in fatty acids and steroids.

FRESHWATER SPONGES AND STEROLS

The 5,6-dihydrostigmasterol (**I**) isolated from *Spongilla lacustris* in 1941, was probably the first natural product isolated from a freshwater sponge (Mazur 1941). In this study, the sterols of the freshwater sponge were separated by repeated adsorption chromatography on alumina. The fractionation allowed the researchers to obtain two steroidal fractions. The more strongly adsorbed fraction resulted in yield of an impure steroid with an absorption spectrum resembling to ergosterol, whilst the other fraction obtained was pure 5,6-dihydrostigmasterol (**I**) (Figure 1).

A characteristic of this kind of apolar extract is that it is not usually able to crystallize as a unique majority substance, as they are commonly composed by complex mixtures.

A study aimed at evaluating the efficiency of mass-analyzed kinetic energy spectrometry (MIKE) of complex mixtures used steroidal fractions of five marine invertebrates and aside freshwater *S. lacustris*, the same species that was evaluated by Mazur (Maquestiau et al. 1978). For its time, before the development of high resolution gas chromatography, the MIKE technique showed good efficiency and allowed the identification of the sterols in mixture. This study enabled the identification of six sterols (Figure 2): (22*E*)-cholesta-5,22-dien-3β-ol (**II**), cholest-5-en-3β-ol (**III**), (24*R*)-24-methyl-cholesta-5,22-dien-3β-ol (**IV**), 24ξ-methylcholest-5-en-3β-ol (**V**), (24*S*)-24-ehylcholesta-5,22-dien-3β-ol (**VI**) and (24*R*)-24-ethylcholest-5-en-3β-ol (**VII**). Maquestiau and collaborators also detected the presence of three other sterols they were unable to identify.

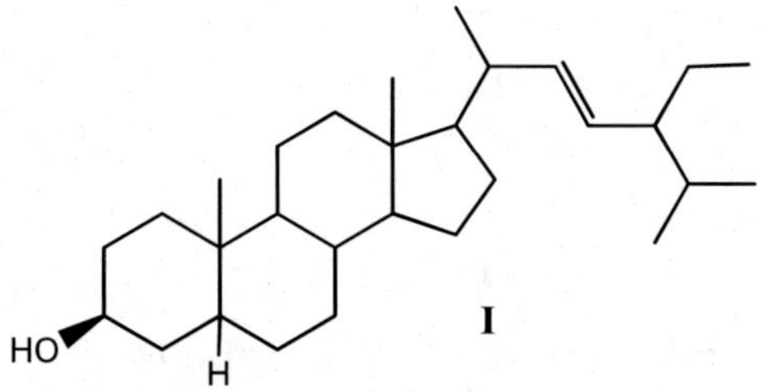

Figure 1. The first natural product isolated from a freshwater sponge: 5,6-dihydrostigmasterol (I).

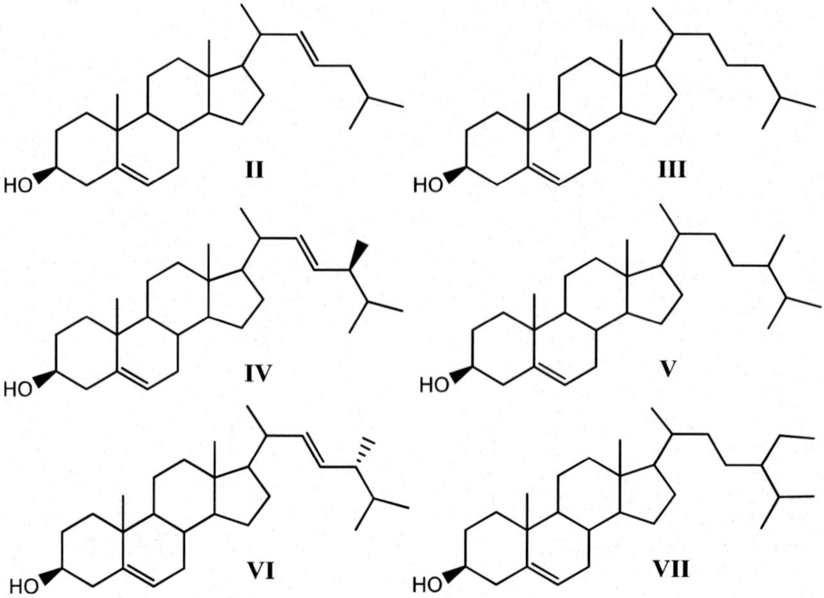

Figure 2. Sterols identified using MIKE.

In 1988, a paper that explored the steroids of two species of freshwater sponges was published (Manconi, Piccialli and Sica 1988). One of the species studied was, once again the *S. lacustris*, and the second species was *Ephydatia fluviatilis*. Sponge extracts were prepared with chloroform-methanol mixture (1:1) and chromatographed with $CHCl_3$ over silica gel providing a steroid fraction. This fraction was acetylated and fractionated by silver nitrate silica gel thin layer chromatography (TLC) followed by high resolution liquid chromatography (HPLC). The resulting purified acetylated sterols were then identified by nuclear magnetic resonance (NMR). In *S. lacustris*, the presence of sterols **II**, **III** and **IV** was confirmed. Additionally, six sterols, (24*R*)-24-methyl-cholesta-5-en-3β-ol (**VIII**), (24*S*)-24-methyl-cholesta-5-en-3β-ol (**IX**), (22*E*,24*S*)-24-methyl-cholesta-5,22-dien-3β-ol (**X**), 24-methyl-cholesta-5,24(28)-dien-3β-ol (**XI**), (24*S*)-24-ethyl-cholest-5-en-3β-ol (**XII**) and (22*E*,24*R*)-24-ethyl-cholesta-5,22-dien-3β-ol (**XIII**), were identified (Figure 3).

Sterol **V**, identified in the Maquestiau study, was not presented with its full absolute configuration established.

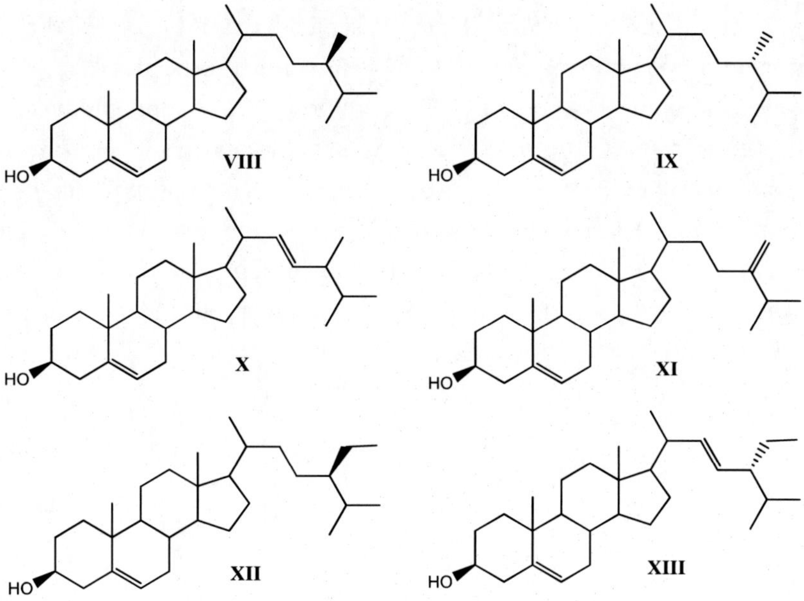

Figure 3. Sterols isolated from *Spongilla lacustris*.

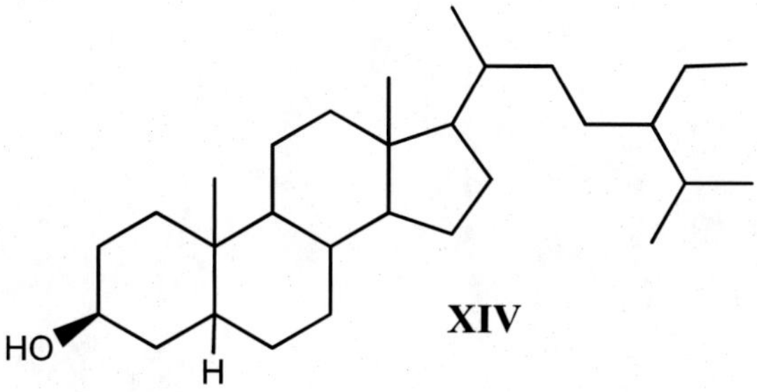

Figure 4. Sterol from *Ephydatia fluviatilis*.

Despite this, Manconi documented the presence of its two enantiomers (**VIII** and **IX**). Similarly, the steroid **IV**, reported by Maquestiau, appeared in the Manconi study. Furthermore, the steroids **XII** and **XIII** are enantiomers of steroids **VII** and **VI**, previously described by Maquestiau. The same sterols reported for *S. lacustris* were found in *E. fluviatilis*, with an additional small amount of the sterol 24-methyl-5α-cholestan-3β-ol (**XIV**) in the *E. fluviatilis* (Figure 4).

In both species, cholesterol **(III)** represented a majority, 74% and 66%, of the steroid fraction in *S. lacustris* and *E. fluviatilis*, respectively.

These initial works, before 1990, exploring the natural products of sponges were restricted to the species, *S. lacustris* and *E. fluviatilis*, most likely because they are the most abundant species in Europe, where the studies were carried out. In 2009, a work from China exploring the lipids of *S. lacustris*, which is used in traditional Chinese medicine, was published (Hu et al. 2009). Hu and co-workers produced an ethanolic extract from the freshwater sponge.

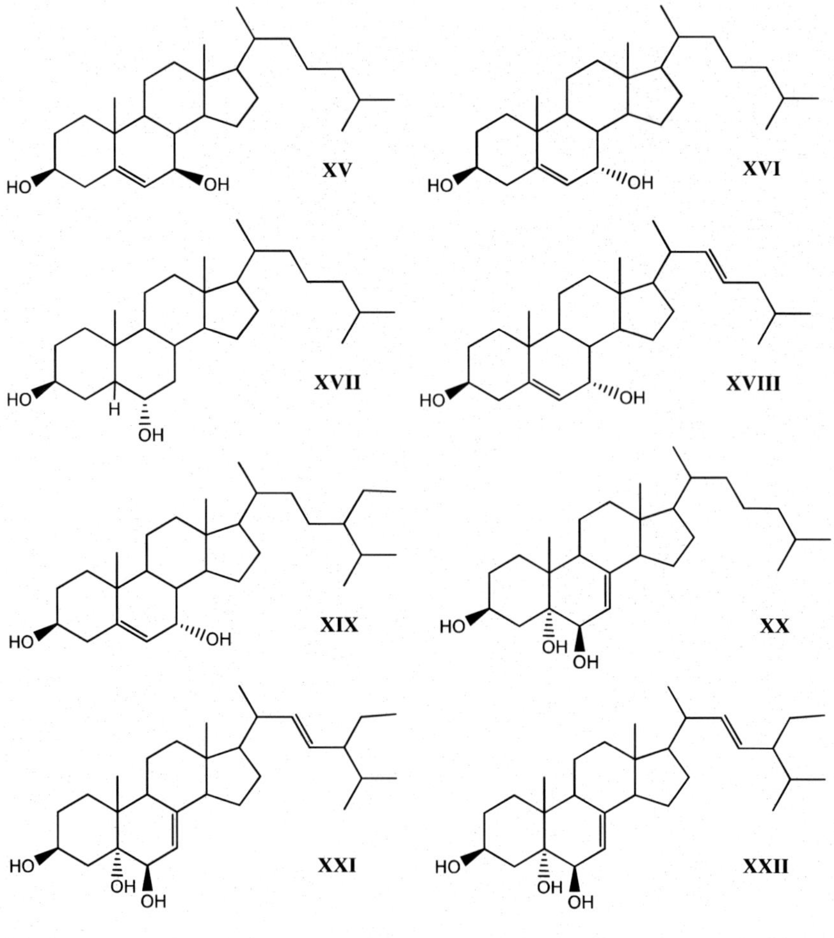

Figure 5. Sterols identified from Chinese *S. lacustris*.

This extract was resuspended and partitioned with organic solvents of increasing polarity. The ethyl acetate soluble fraction was further fractionated with different chromatographic techniques, such as open column chromatography in silica and Sephadex LH-20, recrystallization and HPLC. This led to the identification of the sterols: I, and some polyunsaturated sterols: cholest-5-ene-3β,7β-diol (**XV**), cholest-5-ene-3β,7α-diol (**XVI**), 5α-cholest-7-ene-3β,6α-diol (**XVII**), (22E)-cholest-5,22-diene-3β,7α-diol (**XVIII**), 24ξ-ethylcholest-5-ene-3β,7α-diol (**XIX**), cholest-7-ene-3β,5α,6β-triol (**XX**), (24S)-24-ethyl-cholest-7,22-diene-3β,5α,6β-triol (**XXI**) and (24S)-24-ethyl-cholest-7-ene-3β,5α,6β-triol (**XXII**).

A series of studies with freshwater sponges was also conducted in Russia, exploring endemic species of Lake Baikal, the deepest lake on earth, which contains about 20% of world's freshwater. The first sponge of Lake Baikal to be investigated was *Baicalospongia bacilifera* (Makarieva et al. 1991). An ethanolic extract of the *B. bacilifera* was solubilized in a water-ethanol mixture then partitioned with organic solvents. The hexane soluble fraction was chromatographed over silica gel and Sephadex LH-20 to obtain a sterol fraction. The sterol mixture was subjected to HPLC, resulting in giving the isolated sterols. A new sterol, 24-ethyl-26-norcholesta-5,22E,25-triene-3β-ol (**XXIII**), also called baikolesterol, was identified by NMR (Figure 6). Additionally, (24S,22E)-24-methyl-cholesta-5,22-dien-3β-ol (**XXIV**) was reported for the first time in a freshwater sponge. Sterols **II**, **III**, **IX**, **XI** and **XII** were also detected in the *B. bacilifera*. Another two sterols did not have the absolute configuration determined, and their identification remains in doubt, between sterols **VI** and **X**, or sterols **VI** and **XIII**.

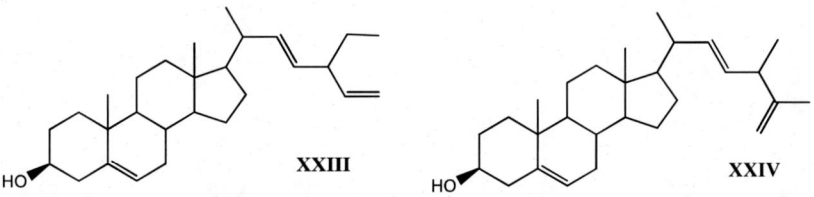

Figure 6. Sterols from Baicalospongia *bacilifera*.

Another Lake Baikal endemic species, *Lubomirskia baicalensis* (Kolesnikova, Makarieva and Stonik 1992), had its steroidal fraction obtained by a method as that reported for *B. bacilifera*. This study reported the isolation of steroids by chromatographic techniques with the steroid fraction being acetylated and then the complex mixture of acetylated steroids analyzed by gas chromatography coupled with mass spectrometry (CG-MS). The identification of the steroids was performed by NMR for those that were isolated, and by retention time and mass spectra for the acetylated steroidal fraction. A further five steroids were newly identified for a freshwater sponge: 24-norcholesta-5,22-dien-3β-ol (**XXV**), 24-methyl-cholesta-5,25-dien-3β-ol (**XXVI**), 24-ethyl-cholesta-5,25-dien-3β-ol (**XXVII**), 24-ethyl-cholesta-5,24*E*(28)-dien-3β-ol (**XXVIII**), 24-ethylcholesta-5,24*Z*(28)-dien-3β-ol (**XXIX**). Other sterols already reported for freshwater sponges that were also detected in *L. baicalensis* were: **II**, **III**, **V**, **XI** and **XXIII**.

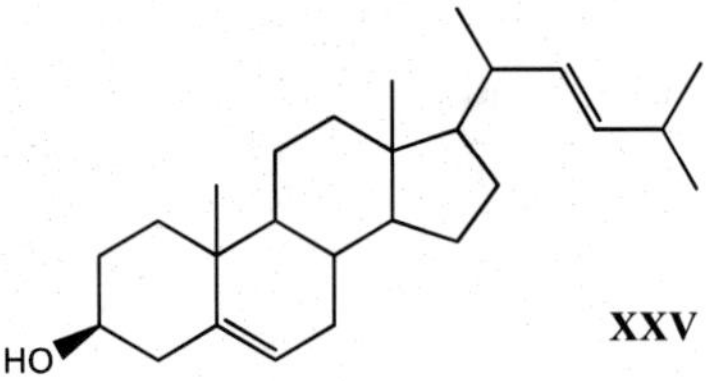

XXV

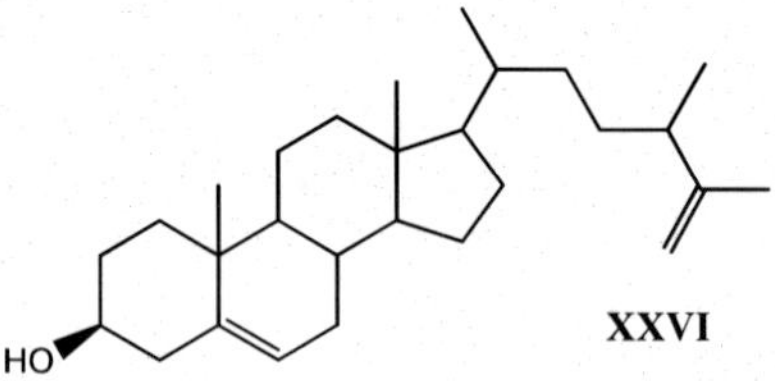

XXVI

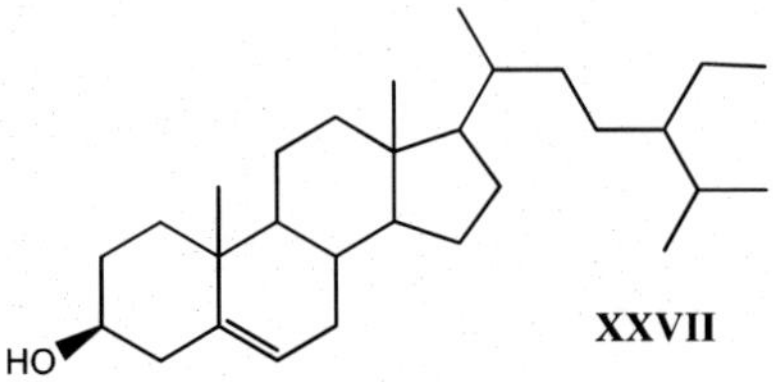

XXVII

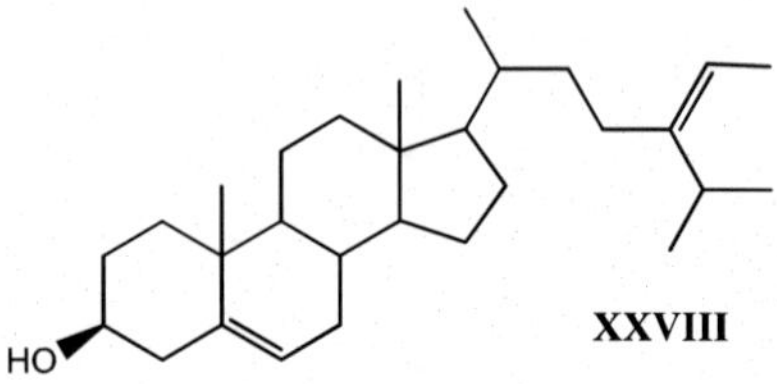

XXVIII

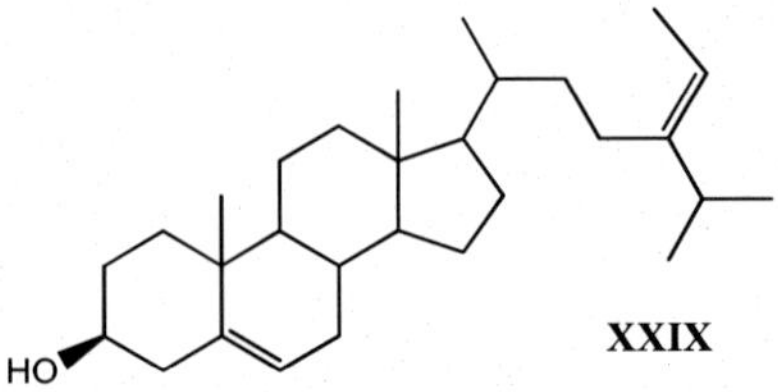

XXIX

Figure 7. Sterols from *Lubomirskia baicalensis*.

For three sterols, the absolute configuration was not established, leaving doubt as to the identity of each between a) **IV** or **X**, b) **VI** or **XIII** and c) **VII** or **XII**. A final sterol, which did not have the stereochemistry determined, was probably **XXIV** or its enantiomer.

More recently, another set of studies have involved sponge collected in the Amazon region.

The first one, from a hexane extract, obtained a steroid fraction by adsorption chromatography techniques, and performed the identification of the steroids by GC-MS comparing retention times and mass spectra (de Barros, Volkmer-Ribeiro and da Veiga Junior 2013). The species evaluated were *Metania reticulata*, *Drulia browni* and *Drulia uruguayensis*, and in the three species the five sterols were identified: **III**, **V**, **IV** and/or **X**, **VI** and/or **XIII,** and **VII** and/or **XII**.

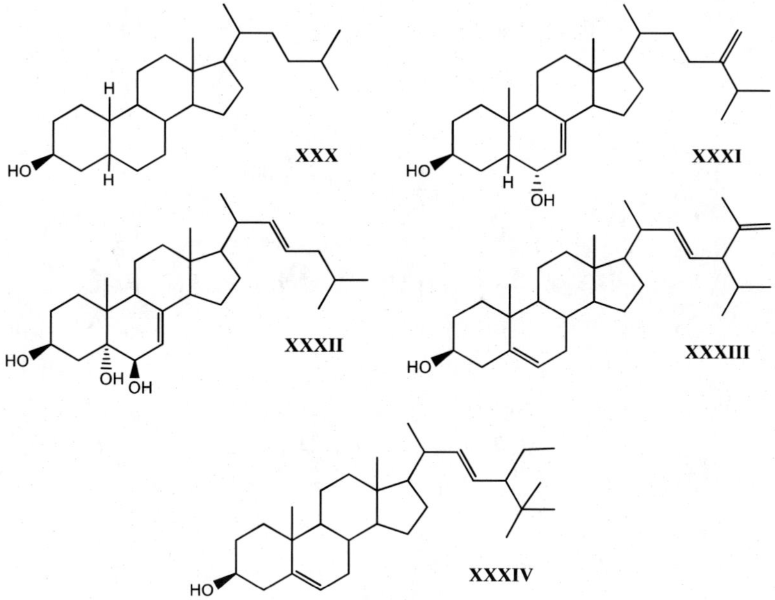

Figure 8. Sterols from Amazonian freshwater sponges.

The major sterol in these Amazon freshwater sponges was 24-ethyl-cholest-5,22-dien-3β-ol (**VI** and/or **XIII**), and not cholesterol, as previously observed in the European and Asian freshwater sponge species.

Another work with the same three Amazonian freshwater sponge species analyzed a methanolic extract of the sponges by electrospray ionization mass spectrometry (Bolson et al. 2019). This technique revealed the presence of five more sterols: 19-nor-5α-cholestan-3β-ol (**XXX**), 24-methylene-5α-cholest-7-en-3β,6α-diol (**XXXI**), cholesta-7,22-dien-3β,5α,6β-triol (**XXXII**), 24-isopropenylcholesta-5,22-dien-3β-ol (**XXXIII**), and 24-ethyl-25-methylcholesta-5,22-dien-3β-ol (**XXXIV**). The sterols **IV**, **XIII** and **XX** are also reported in this study.

Another species of sponge collected in the Amazon region that had its steroidal fraction analysed was *Trochospongilla paulula* (de Barros, Volkmer-Ribeiro and da Veiga Junior 2015). This study was conducted similarly to that reported in the previous work of de Barros and collaborators (2013). In *T. paulula*, the sterols **III**, **V**, **IV** and/or **X**, **VI** and/or **XIII**, and **VII** and/or **XII** were described. With the predominant sterol was also 24-ethyl-cholest-5,22-dien-3β-ol (**VI** and/or **XIII**), as found for the other species of the region.

REFERENCES

Blunt, J. W., Copp, B. R., Keyzers, R. A., Munro, M. H., Prinsep, M. R. (2015). Marine Natural Products. *Natural Products Reports,* v. 32, p. 116 - 211. doi: 10.1039/C4NP00144C.

Blunt, J. W., Copp, B. R., Keyzers, R. A., Munro, M. H. G., Prinsep, M. R. (2016). Marine Natural Products. *Natural Products Reports,* v. 33, p. 382 - 431. doi: 10.1039/C5NP00156K.

Blunt, J. W., Copp, B. R., Keyzers, R. A., Munro, M. H. G., Prinsep, M. R. (2017). Marine Natural Products. *Natural Products Reports*, v. 34, p. 235 - 294. doi: 10.1039/C6NP00124F.

Blunt. J. W., Carroll, A. R., Copp, B. R., Davis, R. A., Keyzers, R. A. Prinsep, M. R. (2018). Marine Natural Products. *Natural Products Reports,* v. 35, p. 8 - 53. doi: 10.1039/C7NP00052A.

Bolson, G. C. M. da C., de Barros, I. B., Volkmer-Ribeiro, C., Lima, J. A., França, T. C. C., Santos, I., Orlandi, P. P., da Veiga Junior, V. F.

(2019). Chemical composition and biological activities of *Metania* and *Drulia* (Metanidae) Freshwater sponges from Amazonia. *Chemistry and Biodiversity*, v. 16, e1900318. doi: 10.1002/ cbdv.201900318.

Burgoyne, D. L., Andersen, R. J., Allen, T. M. (1992). Contignasterol, a highly oxygenated steroid with the unnatural 14β configuration from the marine sponge Petrosia contignata Thiele, 1899. *The Journal of Organic Chemistry*, v. 57, p. 525 - 528. doi: 10.1021/ jo00028a024.

Carrol, A. R., Copp, B. R., Davis, R. A., Keyzers, R. A., Prinsep, M. R. (2019). Marine Natural Products. *Natural Products Reports*, v. 36, p. 122 - 173. doi: 10.1039/C8NP00092A.

Cheung, R. C. F., Ng, T. B., Wong, J. H. Chen, Y. Chan, W. Y. (2016). Marine natural products with anti-inflamatory activity. *Appl. Microbiol. Biotechnol.*, v. 100, p. 1645 - 1666. doi: 10.1007/ s00253-015-7244-3.

de Barros, I. B., Volkmer-Ribeiro, C., da Veiga Junior, V. F. (2013). Sterols from sponges of Anavilhanas. *Biochemical Systematics and Ecology*, v. 49, p. 167 - 171. doi: 10.1016/j.bse.2013.03.022

de Barros, I. B., Volkmer-Ribeiro, C., da Veiga Junior, V. F. (2015). Sterols and volatile compounds of the freshwater sponge *Trochospongilla paulula* (Bowerbank, 1863). *Química Nova*, v. 38, n. 9, p. 1192 - 1195. doi: 10.5935/0100-4042.20150134.

de Goeji, J. M., van den Berg, H., van Oostveen, M. M., Epping, E. H. G., van Duyl, F. (2008). Major bulk dissolved organic carbon (DOC) removal by encrusting coral reef cavity sponge. *Marine Ecology Progress Series*, v. 357, p; 139 - 151. doi: 10.3354/meps07403.

Dembitsky, V. M., Rezanka, T., Srebnik, M. (2003). Lipid compounds of freshwater sponges: family Spongillidae, class Demospongiae. *Chem. Phys. Lipids*, V. 123. p. 117 - 155. doi: 10.1016/S0009-3084(03)00020-3.

Djerassi, C., Silva, C. J. (1991). Sponge sterols: Origin and biosynthesis. *Accounts of Chemical Research*, v. 24, p. 371 - 378.

Gallimore, W. A., Cabral, C., Kelly, M., Scheuer, P. J. (2008). A novel D-ring unsaturated A-nor sterol from the Indonesian sponge, *Axinella carteri* Dendy. *Nat. Prod. Res.,* v. 22, p. 1339 - 1343. doi: 10.1080/14786410601132279.

Goad, L. J. (1983). Steroid biochemistry of marine invertebrates. *Marine Chemistry,* v. 12, p. 225.

Gross, H., König, G. (2006). Terpenoids from Marine Organisms: Unique Structures and their Pharmacological Potential. *Phytochemistry Reviews,* v. 5, p. 115 - 141.

Hickman, J. R. C. P., Roberts, L. S., Larson, A. (2000). *Integrated principles of zoology.* 11 ed. Nova York: McGraw-Hill Science. 928.

Hu, J. M., Zhao, Y.-X., Chen, J.-J., Miao, Z.-H., Zhou, J. (2009). A new spongilipid from the freshwater sponge *Spongilla lacustris. Bulletin of the Korean Chemical Society,* v. 30, p. 1170 - 1172.

Jiménez, C. (2018). Marine Natural Products in Medicinal Chemistry. *ACS Medicinal Chemistry Letters,* v. 9, p. 959 - 961. doi: 10.1021/acsmedchemlett.8b00368.

Kerr, R. G., Baker, B. J., Kerr, S. L., Djerassi, C. (1990). Biosynthetic studies of marine lipids-XXIX. Demonstration of sterol side chain dealkylation using cell-free extracts of marine sponges. *Tetrahedron Letters,* v. 31, p. 5425 - 5428.

Kerr, R. G., Kerr, S. L., Malik, S., Djerassi, C. (1992). Biosynthetic studies of marine lipids. 38.1 mechanism and scope of sterol side chain dealkylation in sponges: Evidence for concurrent alkylation and dealkylation. *Journal of the American Chemical Society,* v. 114, p. 299 - 303.

Kolesnikova, I. A., Makarieva, T. N., Stonik, V. A. (1992). Natural products from the Lake Baikal organisms - II. Sterols from the sponge *Lubomirskia baicalensis. Comparative Biochemistry and Physiology - B Biochemistry and Molecular Biology,* v. 103, p. 501 - 503. doi: 10.1016/0305-0491(92)90326-M.

Makarieva, T. N., Bondarenko, I. A., Dmitrenok, A. S., Boguslavsky, V. M., Stonik, V. A., Chermih, V. I., Efremova, S. M. (1991). Natural products from lake Baikal organisms, I. Baikalosterol, a novel steroid with an unusual side chain, and other metabolites from the sponge

Baicalospongia bacilifera. Journal of Natural Products, v. 54, p. 953 - 958. doi: 10.1021/np50076a005.

Malik, S., Kerr, R. G., Djerassi, C. (1988). Biosynthesis of marine lipids. 19. Dealkylation of the sterol side chain in sponges. *Journal of the American Chemical Society,* v. 110, p. 6895 - 6897.

Manconi, R., Piccialli, V., Sica, D. (1988). Steroids in porifera, sterols from freshwater sponges Ephydatia fluviatilis (L.) and Spongilla lacustris (L.). *Comparative Biochemistry and Physiology -- Part B: Biochemistry and,* v. 91, p. 237 - 245. doi: 10.1016/0305-0491(88)90138-1.

Manconi, R., Pronzato, R. (2008). Global diversity of sponges (Porifera: Spongillina) in freshwater. *Hydrobiologia,* v. 595, p. 27 - 33. doi:10.1007/s10750-007-9000-x.

Maquestiau, A., van Haverbeke, Y., Flammang, R, Mispreuve, H. (1978). Study of complex mariena sterol mixtures by mass-analysed ion kinetic energy spectrometry, *Steroids,* v. 2121, p. 31 - 48.

Mazur, A. (1941). 5,6-Dihydrostigmasterol. *Journal of the American Chemical Society,* v. 63, p. 2442 - 2444.

Mehbub, M. F., Lei, J., Franco, C., Zhang, W. (2014). Marine Sponge Derived Natural products between 2001 and 2010: Trends and Opportunities for discovery of Bioactives. *Marine drugs,* v. 12, p. 4539-4577. doi: 10.3390/md12084539.

Pereira, F. (2019). Have marine natural product drug discovery efforts been productive and how can we improve their efficiency? *Expert Opinion on Drug Discovery,* v. 14, p. 717 - 722. doi: 10.1080/17460441.2019.1604675.

Ruppert, E. E., Fox, R. S., Barnes, R. D. (1996). *Zoologia dos Invertebrados.* 6 ed. São Paulo-SP: Roca, p. 1088.

Van Soest, R. W. M., Boury-Esnault, N., Vacelet, J., Dohrmann, M., Erpenbeck, D., De Voogd, N. J., Santodomingo, N., Vanhoorne, B., Kelly, M., Hooper, J. N. A. (2012). Global Diversity of Sponges (Porifera). *PLoS ONE,* v. 7, e35105. doi: 10.1371/journal.pone.0035105.

Van Soest, R. W. M., Boury-Esnault, N., Hooper, J. N. A., Rützler, K., de Voogd, N. J., Alvarez, B., Hajdu, E., Pisera, A. B., Manconi, R., Schönberg, C., Klautau, M., Kelly, M., Vacelet, J., Dohrmann, M., Díaz, M.-C., Cárdenas, P., Carballo, J. L., Ríos, P., Downey, R., Morrow, C. C. (2019). *World Porifera Database*. Accessed at http://www.marinespecies.org/porifera on 2019-09-27. doi:10.14284/359.

In: Sterols: Types, Classification and Structure ISBN: 978-1-53617-231-7
Editor: Scott Jimenez © 2020 Nova Science Publishers, Inc.

Chapter 3

CULTIVAR EFFECT ON STEROL COMPOSITION OF VIRGIN OLIVE OIL

Bechir Baccouri[1,2,], Hedia Manai-Djebali and Leila Abaza*
[1]Centre of Biotechnology of Borj Cédria,
Laboratory of Olive Biotechnology, Hammam-Lif, Tunisia
[2]University of Tunis El Manar, Tunis, Tunisia

ABSTRACT

Olive oil is obtained from the fruit of several cultivars of olive tree (Olea europea L.), with particular characteristics. Each one of these cultivars exhibits specific physical and biochemical characteristics, providing oils with different compositions and performances.

In olive oil, sterols constitute the majority of the unsaponifiable fraction. In recent years there has been increased interest in the sterols of olive oil for their health benefits and their importance to virgin olive oil (VOO) quality. Several factors are known to affect the sterolic profiles of olive oil. Among these factors is the nature of the cultivar. Recently, it has also been proposed that these profiles could be used to classify virgin

---

[*] Corresponding Author E-mail: bechirbaccouri@yahoo.fr.

olive oils according to their fruit variety. The main sterols found in olive oils were β-sitosterol, Δ5-avenasterol, campesterol and stigmasterol. Cholesterol, 24-methylenecholesterol, clerosterol, campestanol, sitostanol, Δ7-stigmastenol, Δ5,24-stigmastadienol, and Δ7-avenasterol were also found. Most of these compounds are significantly affected by the cultivar type. Chemical characterization of monovarietal olive oil is imperative for the selection of quality cultivars that produce virgin olive oil with good quality. The sterol fraction can be considered as a useful tool to characterize and discriminate monovarietal VOOs. Thus, this chapter is devoted to recent findings concerning the effect of cultivar on sterolic profile of extra virgin olive oil.

Keywords: olive variety, virgin olive oil, sterol, discrimination

INTRODUCTION

This appreciated edible oil is made up of triglycerides (more than 98%) and other minor compounds (about 1-2%) (Alves et al., 2008), such as phenols, tocopherols or sterols. Among this chemical composition, sterols, also called phytosterols, present the greatest proportion of the non-saponifiable fraction of olive oil (Aparicio et al., 2000).

The control of sterol fraction is an important issue for vegetable oil genuineness. Since several studies have demonstrated that each oily fruit has its characteristic sterol profile, its determination could provide abundant information about oil quality (Arafat et al., 2016) and, therefore, it could be used for authentication purposes. Regarding olive oil, sterol composition has been used for the detection of fraudulent admixtures of olive oil with other cheaper vegetable oils, such as hazelnut, corn, soybean, sunflower and cotton seed oils (Boskou, 1996). In addition, this profile has permitted the characterization of olive oils according to their genetic variety and quality grade (virgin, refined, solvent extracted, olive pomace oil, crude olive pomace oil, refined olive-pomace oil, and refined seed oil) (Boskou, 1996). However, composition and total sterol contents are strongly influenced by many factors, such as geographical area (Ben Temime et al., 2008), crop season (Mailer et al., 2010) agronomic practices, such as irrigation (Boskou, 1996), soil conditions (Mailer et al., 2010), storage time and temperature (Boskou, 1996). However, many researchers suggested that cultivar (Baccouri et al., 2018) and fruit maturity (Lazzez et al., 2008) are the two most important factors.

Processing practices particularly affect stigmasterol while horticultural practices and fruit characteristics tend to affect more significantly other sterols such as β-sitosterol, sitostanol, Δ-5-avenasterol and Δ-7-avenasterol (Guillaume et al., 2012).

Spanish (Rivera del Álamo et al., 2004), Italian (Marini et al., 2004), Portuguese (Matos et al., 2007), Tunisian (Baccouri et al., 2018), Turkish (Matthäus and Özcan, 2011), Australian (Guillaume et al., 2012), Greek (Vekiari et al., 2010) and Croatian (Lukić et al., 2013) olive varieties have been evaluated for their sterol composition. All these previous works have demonstrated that sterol profile and content was significantly affected by the olive oil variety.

The study of Guillaume and coworkers clearly demonstrates the strong influence of the variety on sterol composition, particularly in the case of certain sterols such as campesterol, stigmasterol, β-sitosterol and total sterols (Guillaume et al., 2012). Several successful attempts have been made to characterize and differentiate olive oils according to varietal origin on the basis of sterols. High abundance of β-sitosterol, and significant differences between its concentration ranges in oils from different varieties, presumably ensured its robustness as an indicator of varietal origin. Interestingly, similar results were obtained by several studies. The most contributor sterols to classify olive oil samples according to the cultivar are: Δ-7-stigmastenol, 24-methylencholesterol, clerosterol, campesterol, Δ-7-campesterol, β-sitosterol and apparent β-sitosterol (Sanchez-Casas et al., 2004).

A new parameters, the 24-methylene-cholesterol/stigmasterol and Δ-7-campesterol/Δ-5,24-stigmastadienol ratios emerged as reliable variety indicators, these two ratios contributing significantly to the variety differentiation (Lukić et al., 2013). Sterol composition is important for assessing authenticity, detecting adulteration, and characterization purposes because it is specific for each oil. Olive oils command a premium price in the market, leading to great temptation to adulterate by blending them with other lower grade oils such as refined olive oils, vegetable and seed oils.

Results of Yorulmaz and colleagues (2014) revealed that sterol composition was more evident than fatty acid and triacylglycerol profiles for the discrimination of olive varieties. Therefore, they can be used for proper labeling and marketing of varietal oils.

It is well-known that sterol composition can be used to identify adulteration of olive oil, and it has been suggested that it may be used to classify virgin olive oils according to their fruit variety (Lazzez et al., 2008) and to discriminate between different vegetable oils. The study of Lerma-García and coworkers (2008) demonstated that sterol profiles were able to perfectly classify samples belonging to eight different botanical origins (hazelnut, sunflower, corn, olive, soybean, avocado, peanut and grapeseed). The feasibility of classifying extra virgin olive oil (EVOO) according to their genetic variety using the sterol profiles has been demonstrated. EVOO mainly produced at La Comunitat Valenciana (Spain) can be correctly classified with a high reliability by the sequential application of two LDA models, a method proposed by Lerma-García and colleagues (2011).

This method could be helpful for controlling the genetic origin of the olives used to obtain the EVOO (Lerma-García et al., 2008). It was very interesting for olive oil industry; since it provides a fast technique to establish the genetic variety of EVOO. This method could be easily extended to the classification of EVOO obtained from other genetic varieties coming from different areas by constructing new LDA models on the basis of sterol profiles (Lerma-García et al., 2011).

The investigation of Lukić and coworkers (2013) demonstrated that sterols and triterpene dialcohols could be used as reliable indicators of variety and ripening degree among virgin olive oils from three Croatian varieties (Buža, Črna and Rosinjola). Multivariate statistical methods such as PCA and LDA provided efficient differentiation models based on sterols and triterpene dialcohols data. Discriminant analysis proved to be effective in discriminating between olive varieties when it was applied to the sterol data set. Among all sterols the stigmasterol could allow the segregation between varieties (López-Cortés et al., 2013).

Matos and coworkers evaluated the usefulness of three chemical parameters (compositions on tocopherols, sterols and fatty acids) as a tool to discriminate three Portuguese olive oil varieties (Cvs. Cobrançosa, Madural and Verdeal Transmontana). These authors found that the three cultivars were clearly discriminated and campesterol, stigmasterol, clerosterol, β-sitosterol, Δ-5-avenasterol, Δ-7-avenastenol are among the variables that contributed to the discrimination between olive varieties (Matos et al., 2007).

Ranalli and colleagues (2002) studied the contents of sterols, triterpene dialcohols and aliphatic alcohols in oils from seed (non-woody part of the kernel), pulp (mesocarp plus epicarp) and whole olive fruit from seven major Italian varieties. These authors found that the olive varieties or fruit oil kinds appeared to be effectively discriminated through the use of several multivariate methods applied to the data of their sterol or alcohol composition. In particular, PCA and HCA were very effective methods for discriminating the olive variety, whereas canonical discriminant analysis was a very effective technique for differentiating either the olive variety or the fruit oil kind.

Boulkroune and colleagues (2017) characterized the sterolic and alcoholic fractions of four Algerian olive oil varieties (Chemlal, Mekki, Aghenfas and Buichret), they reported that the most discriminating variables were cholesterol, campestanol, Δ-7-stigmastenol, Δ-7-avenasterol, clerosterol, alcohol C22, alcohol C25, alcohol C27.

Only few works have been published regarding sterol composition of minor Tunisian cultivars because most research studies related to the Tunisian olive oil have been focused on the characterization of the main variety (Chemlali). This fact causes a lack of information of the chemical composition of several minor cultivars that could provide a better oil quality.

As previously reported by other authors [25, 31, 32], the composition of sterols in olive oil varied according to the cultivar and maturity index. In general, the most abundant sterol was β-sitosterol; its relative content was comprised between 71 and 89%, followed by Δ7 + Δ5-avenasterol, which ranged between 4 and 23%. The content of β-sitosterol generally decreases during ripening, while Δ7 + Δ5-avenasterol content increases, which could be explained by the presence of desaturase activity [31, 33]. The highest β-sitosterol content (88.72%) and the lowest one for Δ7 + Δ5-avenasterol (4.59%) were found for Zarrazi cultivar, whereas Fouji and Dokhar are characterized by the lowest content of β-sitosterol (71.02 and 73.44%) and the highest of Δ7 + Δ5-avenasterol (21.18 and 22.83%). These values are similar to those reported for other olive oil cultivars [24, 25, 31, 32].

The other main phytosterols identified in these extra virgin olive oils are stigmasterol and campesterol. Their contents vary from one cultivar to another. Stigmasterol is present in all samples in lesser amounts than campesterol, which indicates that all oil samples have been obtained from healthy fruits, naturally ripened on the plant [10]. Also the Campesterol never exceed the upper limit established by International Olive Oil Council (4%). The campesterol content for Picholine marocaine and Arbequina was significantly higher than the values of the other cultivars (66.36 and 71.04 mg kg-1, respectively).

METHODS

1. Oil Samples

Olives were handpicked in perfect sanitary conditions. The selected olives were picked at the same stage of ripeness, and their oils were extracted with the same processing system. After harvesting, the olive fruit samples were immediately transported to the laboratory mill, where the oils were extracted using an Abencor analyzer. Olives (1.5-2 kg) were crushed with a hammer mill and slowly mixed for 30 min, centrifuged without addition of warm water, and then transferred into dark glass bottles. All samples were stored at 40C in darkness in amber glass bottles until analysis.

2. Analytical Determinations

2.1. Determination of Sterols

The qualitative and quantitative sterol contents of the samples were determined according to the European Official Analysis Methods, described in Annexes V and VI of Regulation EEC/2568/91 of the European Union Commission.

The determination gave sterols expressed in ppm total sterols only and% individual sterols. The oil sample was saponified with ethanolic potassium hydroxide solution. The unsaponifiable fraction was removed with ethyl ether. The unsaponifiable sterol fraction was separated by chromatography on silica gel plate. Separation and quantification of the silanised sterol fraction was carried out by capillary gas chromatography, on a Hewlett Packard 6890 chromatograph with autosampler and flame ionisation detector (FID) using an HP-5MS capillary column (30 m x 0.25 mm x 0.25 µm). The working conditions of the chromatograph were: injector 300°C, isothermal column 260°C, and detector 325°C. The injected quantity was 0.2 µl, at a flow rate of 1.1 ml/min, using helium as carrier gas. Quantification was made by addition of an internal standard (α-cholestanol) and apparent b-sitosterol was calculated as the sum of β-sitosterol, $\Delta 5,23$-stigmastadienol, clerosterol, sitostanol and $\Delta 5,24$-stigmastadienol. Sterols peak identification was carried out according to the reference method. Figure 1 shows a chromatogram obtained for one of the samples.

2.2. Statistical Analysis

The results are reported as the mean values. Data were compared on the basis of standard deviation of the mean values. In addition, Duncan's multiple range tests were used to determine significant differences among data. All collected data were submitted to Hierarchical clustering analysis. The associations obtained based on the similarity in Euclidian distances using the Statistica 5.0 package. Principal components analysis is performed with XLSTAT software, Version 2014.

RESULTS AND DISCUSSION

Sterols, which comprise a major portion of the unsaponifiable matter, are found in almost all fats and oils and are also characteristic of the purity of vegetable oils. The composition of the steroidal fraction of the olive oil is a very useful parameter for detecting adulterations or verifying authenticity, because it can be considered as a distinct fingerprint (Boskou, 1996). Table 1 shows the phytosterol composition of monovarietal virgin olive oils of the studied varieties. The four extra virgin olive oils shows a phytosterol composition in compliance within the established limits, which ranges depend on the varieties ($P < 0.05$).

Table 1. Sterol composition of studied olive oil samples (Results expressed as percentage of total sterols)

	Chetoui	Rekhami	Neb Jmel	Zarrazi	Norm
Cholesterol%	0.13[a]	0.14[a]	0.07c	0.2	≤0.5
24-methylencholesterol%	0.3[a]	0.28[a]	0.133[b]	0.30[a]	
Campesterol%	2.59[b]	2.64[b]	2.88[b]	3.55[a]	≤ 4
Campestanol%	0.08[b]	0.1[a]	0.05[b]	0.11[a]	
Stigmasterol%	1.12[a]	0.64[b]	0.55[b]	0.58[b]	Campesterol
Clerosterol%	0.94[a]	1.06[a]	0.95[a]	1.1[a]	
β-Sitosterol%	78[b]	83[a]	85.3[a]	82.7[a]	
Sitostanol%	0.27[b]	0.48[a]	0.22[b]	0.47a	
Δ5-Avenasterol%	15.4[a]	0.5[c]	8.14[b]	9.55[b]	
Δ5,24 Stigmastadienol%	0.64[a]	0.48[b]	0.68[a]	0.59[b]	
Δ7-Stigmastenol%	0.18[c]	0.34[b]	0.74[a]	0.27[b]	≤0.5
Δ7-Avenasterol%	0,3[b]	10.5[a]	8.14[a]	0.61[b]	
The apparent β Sitosterol	95.82[a]	95.44[a]	95.3[a]	94.38[a]	
Total sterols mg/kg	1350[b]	1650[b]	2040[a]	1100[d]	

a,b,c: Different superscripts for the same quality parameter mean significant ifferences among varieties (n = 6; $p < 0.01$). nd: not detected. *Apparent β-Sitosterol = β-Sitosterol + Δ5-Avenasterol + Clerosterol + Sitostanol + Δ5,24-Stigmastadienol. Results expressed as percentage of total sterols.

Concerning total sterols, all of the monovarietal oils studied contained more than 1000 mg kg-1, the minimum value established by the EU regulation for the "extra virgin" olive oil category (Table 1). Significant differences between cultivars were observed in total sterol contents. Neb Jemal oil showed the highest value (2040 mg kg-1), while Zarazi had the lowest value (1100 mg kg-1).

Results show that the main sterols found in all of the olive oils studied were βsitosterol, Δ5-avenasterol, and campesterol. βSitosterol is the most abundant sterol in olive oil and has a recognized effect on lowering cholesterol levels by opposing the absorption of cholesterol in the intestinal tract. Thus, The lowest β-sitosterol content (88.72%) and the highest one for Δ5-avenasterol (15.4%) were found for Chetoui cultivar, whereas Neb Jemal and Rekhami are characterized by highest content of β-sitosterol (83 and 85%) and the highest of Δ7-Stigmastenol (0.74 and 0.34%), respectively.

Moreover, the campesterol content varied from one cultivar to another, but its level did not exceed the upper limit established by the EU regulation (4%). The percentages of stigmasterol in all tested samples were lower than those of campesterol, which shows that all of the oil samples came from healthy fruits that were not obtained by forcing systems. Thus, the stigmasterol content for Chétoui was significantly higher than the values of the other varieties.

The studied Tunisian cultivars showed very low amounts of campesterol when compared with those from other studied cultivars, namely the Spanish ones studied by Rivera et al. (2004) and the Portuguese ones studied by Alves et al. (2005). The minor components cholesterol and Δ7-stigmastenol, which showed no significant differences among varieties ($p < 0.01$), were below the established limit of 0.5%, although these are minor components with 24-methylenecholesterol and Δ7-avenasterol in the sterol fraction. 24-Methylenecholesterol is an immediate metabolite in the synthesis of campesterol and is characteristic of the oil in the pulp of the olive, but is not found in the oil of the stones (Christopoulou et al., 1996).

The Δ5,24-stigmastadienol content for the Rekhami cultivar (0.48%) was lower than the values for the other cultivars. Cholesterol and Δ7-

stigmastenol were quantified and compared to other authenticity indices established by the current legislation. Hence, in all cultivars, the individual percentages were below the established limit of 0.5%. Therefore, the highest sitostanol content was found in the Rekhami sample.

On the other hand, the apparent βsitosterol, expressed by the sum of the contents of βitosterol and the other four sterols formed by the degradation of βsitosterol (sitostanol, Δ5,24-stigmastadienol, clerosterol, and Δ5-avenasterol), was higher than the minimum established (93%) in all oil samples.

Several researchers suggested the effectiveness of sterol composition on the classification of EVOO according to the variety. Some authors mentioned that statistical analysis applied to only sterol data are sufficient to discriminate EVOO according to their varieties, but others reported that these components are effective in olive variety discrimination when they are used together with other components such as fatty acids, triacylglycerols, aliphatic alcohols, triterpene dialcohols and tocopherols.

The investigation of Lukić and coworkers (2013) demonstrated that sterols could be used as reliable indicators of variety and ripening degree among virgin olive oils from three Croatian varieties (Buža, Črna and Rosinjola). Multivariate statistical methods such as PCA provided efficient differentiation models based on sterols and triterpene dialcohols data. Discriminant analysis proved to be effective in discriminating between olive varieties when it was applied to the sterol data set. Among all sterols the stigmasterol could allow the segregation between varieties (López-Cortés et al., 2013).

Matos and coworkers evaluated the usefulness of three chemical parameters (compositions on tocopherols, sterols and fatty acids) as a tool to discriminate three Portuguese olive oil varieties (Cvs. Cobrançosa, Madural and Verdeal Transmontana). These authors found that the three cultivars were clearly discriminated and campesterol, stigmasterol, clerosterol, β-sitosterol, Δ-5-avenasterol, Δ-7-avenastenol are among the variables that contributed to the discrimination between olive varieties (Matos et al., 2007). It is well-known that sterol composition can be used to identify adulteration of olive oil, and it has recently been suggested that it may be used to classify virgin olive oils according to their fruit variety.

CONCLUSION

It is important to point out that variety is among the most important factors affecting the profile of the sterol fraction in olive oil. The total content of sterols and their composition in olive oil is strongly influenced by genetic factors (variety). The most abundant sterol in olive oil is β-sitosterol, which represents a 75-85% of the total sterol fraction. Other sterols are found in considerable amounts, such as Δ5-avenasterol (0.5-15%) and campesterol (approximately 3% of the total sterol fraction). However, other sterols, such as cholesterol, campestanol, stigmasterol, Δ5,24-stigmastadienol, Δ7-stigmastenol and Δ7-avenasterol, among others, are present in studied olive oil in small amounts.

REFERENCES

Alves, M. R., Cunha, S. C., Amaral, J. S., Pereira, J. A. & Oliveira, M. B. (2005). Classification of PDO olive oils on the basis of their sterol composition by multivariate analysis. *Anal. Chim. Acta*, 549, 166-178.

Alves, M. R., Cunha, S. C., Amaral, J. S., Pereira, J. A. & Oliveira, M. B. (2005). Classification of PDO olive oils on the basis of their sterol composition by multivariate analysis. *Anal. Chim. Acta*, 549, 166-178.

Aparicio, R. & Aparicio-Ruiz, R. (2000). Authentication of vegetable oils by chromatographic techniques. *J. Chromatogr. A*, 881, 93-104.

Arafat, S. M., Basuniy, A. M. M., Elsayed, M. E. & Soliman, H. M. (2016). Effect of pedological, cultivar and climatic condition on sterols and quality indices of olive oil. *Scientia Agric.*, 13(1), 23-29.

Baccouri, B., Zarrouk, W., Guerfel, M., Baccouri, O., Nouairi, I., Krichene, D., Daoud, D. & Zarrouk, M., (2008). Composition, quality and oxidative stability of virgin olive oils from some selected wild olives (Olea europaea L. subsp. Oleaster). *Grasas Aceites* 59 (4), 346-351.

Baccouri, B., Manaia, H., Casas, J. S., Osorio, E. & Zarrouk, M., (2018). Tunisian wild olive (Olea europaea L. subsp. oleaster) oils: Sterolic and triterpenic dialcohol compounds. *Ind. Crops Prod.* 120, 11-15.

Ben Temime, S., Manai, H., Methenni, K., Baccouri, B., Abaza, L., Daoud, D., Sanchez-Casas, J., Osorio-Bueno, E. & Zarrouk, M. (2008). Sterolic composition of Chetoui virgin olive oil: influence of geographical origin. *Food Chem.*, 110, 368-374.

Boskou, D. (1996). Olive oil chemistry and technology. AOCS press, Champaign, pp. 12-52.

Boulkroune, H., Lazzez, A., Guissous, M., Bellik, Y. Smaoui, S., Grati Kamoun N. & Madani, T. (2017). Characterization of sterolic and alcoholic fractions of some Algerian olive oils according to the variety and ripening stage. *OCL-Ol Corps Gras Li.* (doi: 10.1051/ocl/20170 26).

Guillaume, C., Ravetti, L., Ray, D. L. & Johnson, J. (2012). Technological factors affecting sterols in Australian olive oils. *J. Am. Oil Chem. Soc.*, 89, 29-39.

Lazzez, A., Perri, E., Caravita, M. A., Khlif, M. & Cossentini, M. (2008). Influence of olive maturity stage and geographical origin on some minor components in virgin olive oil of the Chemlali variety. *J. Agric. Food Chem.*, 56, 982-988.

Lerma-García, M. J., Ramis-Ramos, G., Herrero-Martínez, J. M. & Simó-Alfonso, E. F. (2008). Classification of vegetable oils according to their botanical origin using sterol profiles established by direct infusion mass spectrometry. *Rapid Commun. Mass Spectrom.*, 22, 973-978.

Lerma-García, M. J., Simó-Alfonso, E. F., Méndez, A., Lliberia, J. L. & Herrero-Martínez, J. M. (2011). Classification of extra virgin olive oils according to their genetic variety using linear discriminant analysis of sterol profiles established by ultra-performance liquid chromatography with mass spectrometry detection. *Food Res. Int.*, 44, 103-108.

López-Cortés, I., Salazar-García, D. C., Velázquez-Martí, B. & Salazar, D. M. (2013). Chemical characterization of traditional varietal olive oils in East of Spain. *Food Res. Int.*, 54, 1934-1940.

Lukić, M., Lukić, I., Krapac, M., Sladonja, B. & Piližota, V. (2013). Sterols and triterpene diols in olive oil as indicators of variety and degree of ripening. *Food Chem.*, 136, 251-258.

Mailer, R. J., Ayton, J. & Graham, K. (2010). The influence of growing region, cultivar, and harvest timing on the diversity of Australian olive oil. *J. Am. Oil Chem. Soc.*, 87, 877-884.

Marini, F., Balestrieri, F., Bucci, R., Magrì, A. D., Magrì, A. L. & Marini, D. (2004). Supervised pattern recognition to authenticate Italian extra virgin olive oil varieties. *Chemom. Intell. Lab. Syst.*, 73, 85-93.

Matos, L. C., Cunha, S. C., Amaral, J. S., Pereira, J. A., Andrade, P. B., Seabra, R. M. & Oliveira, B. P. P. (2007). Chemometric characterization of three varietal olive oils (Cvs. Cobrançosa, Madural and Verdeal Transmontana) extracted from olives with different maturation indices. *Food Chem.*, 102, 406-414.

Matos, L. C., Cunha, S. C., Amaral, J. S., Pereira, J. A., Andrade, P. B., Seabra, R. M. & Oliveira, B. P. P. (2007). Chemometric characterization of three varietal olive oils (Cvs. Cobrançosa, Madural and Verdeal Transmontana) extracted from olives with different maturation indices. *Food Chem.*, 102, 406-414.

Matthäus, B. & Özcan, M. M. (2011). Determination of fatty acid, tocopherol, sterol contents and 1,2- and 1,3-diacylglycerols in four different virgin olive oil. *J. Food Process Technol.*, 2, 1-4.

Ranalli, A., Pollastri, L., Contento, S., Di Loreto, G., Iannucci, E., Lucera, L. & Russi, F. (2002). Sterol and alcohol components of seeds, pulp, and whole olive fruit oils. Their use to characterize olive fruit variety by multivariates. *J. Sci. Food Agric.*, 82, 854-859.

Rivera del Álamo, R. M., Fregapane, G., Aranda, F., Gomez-Alonso, S. & Salvador, M. D. (2004). Sterol and alcohol composition of Cornicabra virgin olive oil: the campesterol content exceeds the upper limit of 4% established by EU regulations. *Food Chem.*, 84, 533-537.

Vekiari, S. A., Oreopoulou, V., Kourkoutas, Y., Kamoun, N., Msallem, M., Psimouli, V.& Arapoglou, D. (2010). Characterization and seasonal variation of the quality of virgin olive of the Throumbolia and Koroneiki varieties from Southern Greece. *Grasas Aceites*, 61, 221-231.

Yorulmaz, A., Yavuz, H. & Tekin, A. (2014). Characterization of Turkish olive oils by triacylglycerol structures and sterol profiles. *J. Am. Oil Chem. Soc.*, 91, 2077-2090.

"Reviewed by Pr. Mokhtar Zarrouk, Centre of Biotechnology of Borj Cédria, Tunisia."

In: Sterols: Types, Classification and Structure   ISBN: 978-1-53617-231-7
Editor: Scott Jimenez   © 2020 Nova Science Publishers, Inc.

Chapter 4

INVESTIGATION OF STEROL COMPOUNDS OF VIRGIN OLIVE OIL FROM NEW CULTIVARS OBTAINED THROUGH UNCONTROLLED CROSSINGS

Bechir Baccouri[1,3] *and Imene Rajhi*[2,3]
[1]Centre of Biotechnology of Borj Cédria,
LR15CBBC05 Laboratory of Olive Biotechnology,
B.P. 901, Hammam-Lif, Tunisia
[2]Centre of Biotechnology of Borj Cédria,
Laboratory of Legumes, Hammam-Lif, Tunisia
[3]University of Tunis El Manar, B.P. 94,
Tunis, Tunisia

ABSTRACT

This work was carried out on the study of virgin olive oil from two new olive varieties obtained through uncontrolled crossings. Preliminary work evaluating the oil fatty acid composition of the oil of 50 descendants showed the performance of two cultivars among the studied hybrids. These two new cultivars (B1 and B2) have an improved oil composition compared to that of Chemlali, the most dominant Tunisian olive oil variety. A further study was therefore required for their complete characterization. In the present study, we proposed to determine the sterolic composition. Considering the percentages of the major sterols identified and quantified in samples, β-sitosterol was the major compound for all oils with percentages of apparent β-sitosterol (sum of β-sitosterol, clerosterol, Δ-5-avenasterol) of 93%. Campesterol and stigmasterol were in all cases very close to the legal limits for olive oil ($\leq$ 4% and < campesterol, respectively). The statistical analysis showed significant differences between oil samples, and the obtained results showed that the great variability of the oil composition between varieties is influenced exclusively by the genetic factor. This study aims to contribute to the optimisation and valorisation of virgin olive oil quality in the world olive-producing areas.

Keywords: Olive oil, New olive variety, Sterol, β-sitosterol, Campesterol

INTRODUCTION

Tunisia is one of the countries in the olive oil producing world. It is the largest African exporter and in the fourth place worldwide after Spain, Italy and Greece (IOOC 2004). Many varieties are cultivated in Tunisia but there are two, which stand out: Chemlali, a cultivar that occupies more than 2/3 of the total olive growing area, and it is cultivated in the centre and in the south of the country. The Chetoui variety, on the other hand, is the second main variety cultivated in Tunisia (Ben Temime et al. 2008). It is widespread in the north of the country, occurring in plains as well in mountain regions (Ben Temime et al. 2006). A major effort has been made recently to improve the quality of the olive oil produced in Tunisia (Baccouri et al. 2011).

Olives exist in two forms, namely wild or oleaster (Olea europaea subsp. europaea var. sylvestris) and cultivated (O. europaea subsp. europaea var. europaea). Cultivated olives propagate vegetatively by cutting or grafting and they can be considered as varieties of unknown origin. However, wild olives only reproduce sexually by wind pollination and their seeds are mainly dispersed by birds (Uncontrolled crossings) (Baccouri et al. 2008). Wild olives include true oleasters (wild forms present in undisturbed areas) and feral forms (resulting from cultivars escaping cultivation or products of hybridization between a variety and a nearby true oleaster and usually found in disturbed or abandoned fields) (Ouazzani et al. 1999). The olive (Olea europaea L.) is a predominantly allogamous species showing a high degree of outcrossing. Most pollen transfer is by wind (Morettini and Pulselli 1953), with insect involvement occurring to a small extent. Progeny are readily derived from crosses between cultivated varieties (cultivars), as well as between cultivars and both feral (escaped) and wild (oleaster) olives (Angiolillo et al. 1999) The olive and the wild olive or oleaster border in many places, but little is known on their relationships, and whether in each place oleaster are feral, i.e., they derived from the olive, or are genuine or natural (Ouazzani et al. 1999) and the source of cultivars.

In Tunisia, the situation is exemplary and since the studies by Camps-Fabrer (1953) many reports have shown the diversity for the Tunisian olive (Baccouri et al. 2008), but little attention has been given to the Tunisian oleaster(Baccouri et al. 2011. The colonisations of Tunisia by Phoenicians, Greeks and Romans make likely the introduction of foreign cultivars. However, the oleaster is indubitably spontaneous in this area, and if local domestication events occurred, we should expect tight relationships between some cultivars and oleasters (Mekuria et al. 1999; Spennemann and Allen 2000). Olive oil naturally contains several minor compounds such as hydrocarbons, sterols, aliphatic alcohols, phenolic compounds, tocopherols, pigments and flavor components. Sterols are the main constituents of the unsaponifiable fraction and their content corresponds to around 20% of the total unsaponifiable matter of olive oil Giacalone et al. 2015; Guillaume, C. & Ravetti et al. 2015).

It is essential that superior wild (feral) material is identified and propagated before it is destroyed, as it represents the progeny of an uncontrolled outcrossing 'experiment'. Based on this wild material, which is clearly well-adapted to Tunisia, there is the potential to increase Tunisian olive oil production.

This is a first study on the sterolic fraction of olive oils obtained from feral olive trees growing in Tunisia to define new cultivars well adapted to Tunisian environment and yielding high quality oils Many publications report on the composition of sterol compounds of monovarietal and oleasters oils (Baccouri et al. 2018), but there are no studies about the mentioned composition of oil obtained from feral olive trees. Although the study was carried out in Tunisia, it might be applied to other countries with wild olive trees, in order to contrast productivity and oil quality.

METHODS

1. Oil Samples

Three feral olive populations originating from different regions of Tunisia were sampled totaling 50 trees. All trees were tagged and their exact location was noted. Later, two feral olives showing the best fatty acid and pomological characteristics were selected and sampled for further chemical analysis. Olives were hand picked in perfect sanitary conditions. The selected feral olives were picked at the same stage of ripeness, and their oils were extracted with the same processing system.

After harvesting, the olive fruit samples were immediately transported to the laboratory mill, where the oils were extracted using an Abencor analyzer. Olives (1.5 – 2kg) were crushed with a hammer mill and slowly mixed for 30 min, centrifuged without addition of warm water, and then transferred into dark glass bottles. All samples were stored at 40C in darkness in amber glass bottles until analysis.

2. Analytical Determinations

2.1. Determination of Sterols

The qualitative and quantitative sterol contents of the samples were determined according to the European Official Analysis Methods, described in Annexes V and VI of Regulation EEC/2568/91 of the European Union Commission.

The determination gave sterols expressed in ppm total sterols only and% individual sterols. The oil sample was saponified with ethanolic potassium hydroxide solution. The unsaponifiable fraction was removed with ethyl ether. The unsaponifiable sterol fraction was separated by chromatography on silica gel plate. Separation and quantification of the silanised sterol fraction was carried out by capillary gas chromatography, on a Hewlett Packard 6890 chromatograph with autosampler and flame ionisation detector (FID) using an HP-5MS capillary column (30m x 0.25mm x 0.25μm). The working conditions of the chromatograph were: injector 300°C, isothermal column 260°C, and detector 325°C. The injected quantity was 0.2μl, at a flow rate of 1.1ml/min, using helium as carrier gas. Quantification was made by addition of an internal standard (α-cholestanol) and apparent b-sitosterol was calculated as the sum of β-sitosterol, Δ 5,23-stigmastadienol, clerosterol, sitostanol and Δ 5,24-stigmastadienol. Sterols peak identification was carried out according to the reference method. Figure 1 shows a chromatogram obtained for one of the samples.

2.2. Statistical Analysis

The results are reported as the mean values. Data were compared on the basis of standard deviation of the mean values. In addition, Duncan's multiple range tests were used to determine significant differences among data. All collected data were submitted to Hierarchical clustering analysis. The associations obtained based on the similarity in Euclidian distances using the Statistica 5.0 package. Principal components analysis is performed with XLSTAT software, Version 2014.

RESULTS AND DISCUSSION

Sterol profiles are used to characterize virgin olive oils and especially to detect the adulteration of olive oil with other oils (Abumweis et al.

2008). It has also been proposed that these profiles can be used to classify virgin olive oils according to their fruit variety.

Sterols are important constituents of olive oils because they relate to the oil quality. Besides, their determination is of major interest due to their health benefits. The analysis of the sterols by gas chromatography on a capillary column indicated the presence of thirteen sterols. Table 1 lists the sterol levels obtained for the different oils. Regarding the authenticity indices established by the European legislation, all the analyzed oils showed a sterolic composition within the established limit for extra virgin olive oil category. In all cases, total sterols were remarkably higher than the minimum limit set by legislation (1000mg/kg). Such a high sterol content is undoubtedly a good characteristic for olive oils since sterols show great benefits for human health. Studied oils showed significant differences ($p < 0.01$) in their sterol levels. B1 showed the highest total sterol content (2000mg/kg), and B2 had the lowest (1600mg/kg). These contents were high compared with the autochthonous oil for Chemlali oil (1700mg/kg) (Figure 1).

These findings are in good agreement with those of other authors working on Tunisian olive oil varieties (Stiti et al. 2002). They reported that the content of total sterols in seven monovarietal Tunisian oils (Chemchali, Chemlali, Chetoui, Gerboui, Ouelati, Sayali and Zalmati) varied between 1082 and 2017mg/kg; these values were low compared with those from Portuguese oils studied by Alves et al. (2005) who found values ranging from 2003 to 2682mg/kg.

The quantitative study of this fraction revealed significant differences in the percentage of sterols among all studied oils. As expected for Chemlali, the main sterols found in all olive oil samples studied were β-sitosterol, Δ5-avenasterol, and campesterol (Table 1). There was a clear predominance of the β-sitosterol, which is a very interesting sterol from a biological view because it opposes the intestinal absorption of the cholesterol. Relative to the authenticity indices established by current legislation (EU Regulation 2016), the campesterol content was remarkably high. These compounds showed changes in the studied oils according to the cultivar.

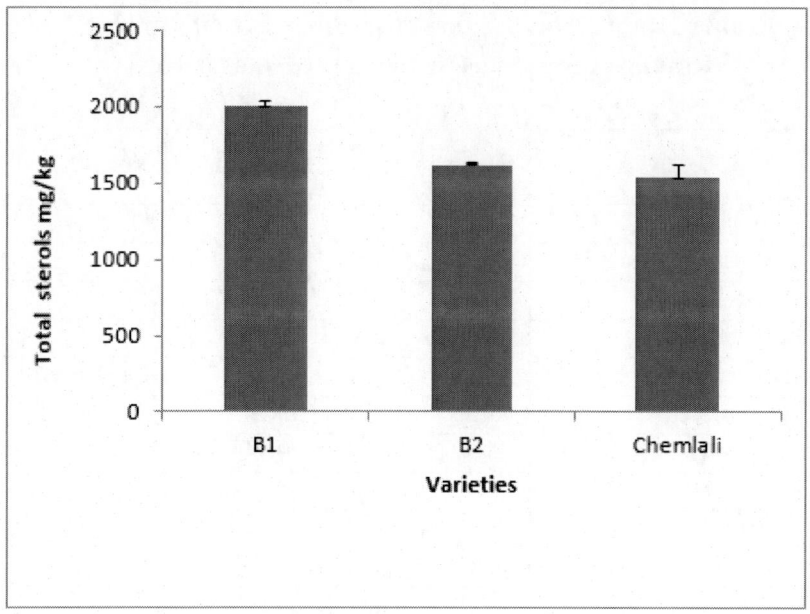

Figure 1. Total Sterols of studied olive oil samples.

In fact, Chemlali oil showed the highest content (3.60mg/kg) whereas B1 oil presented the lowest values (2.22mg/kg); B2 oil had intermediate value. The contents of stigmasterol in the samples were lower than those of campesterol with a mean value of less than 2%. Our data showed that levels of the $\Delta 7$-campesterol, clerosterol or cholesterol contents are not useful for discriminating among the tested oils. Stigmasterol is related to various parameters of the quality of virgin olive oil. High levels correlate with high acidity and low organoleptic quality. The analysed samples presented low levels of stigmasterol, which is indicative that the oil came from healthy fruits and not obtained by systems of forcing.

To benchmark our oils against other authenticity indices established by current legislation, we examined the level of apparent β-sitosterol, which is the sum of the content of β-sitosterol and four chromatographically adjacent phytosterols formed by the degradation of β-sitosterol: sitostanol, $\Delta 5,24$-stigmastadienol, clerosterol, and $\Delta 5$-avenasterol was >93% in all the oils analyzed. This is the regulatory minimum limit, and indicates that the sum of the remaining sterols does not surpass 7%.

Table 1. Sterol composition of studied olive oil samples
(Results expressed as percentage of total sterols)

	B1	B2	Chemlali	Norm
Cholesterol%	0.23[a]	0.12[b]	0.11[b]	≤0.5
24-methylencholesterol%	0.08[b]	0.04[b]	0.18[a]	
Campesterol%	3.55[a]	2.22[b]	3.60[a]	≤ 4
Campestanol%	0.26[a]	0.10[b]	0.05[c]	
Stigmasterol%	1.40[a]	0.68[b]	0.39[c]	Campesterol
Δ 7-Campesterol%	nd	0.12[a]	nd	
Clerosterol%	1.03[a]	1.11[a]	0.94[b]	
β-Sitosterol%	85.59[b]	87.46[a]	85.89[ab]	
Sitostanol%	1.89[a]	0.91[b]	0.38[c]	
Δ5-Avenasterol%	4.30[c]	5.79[b]	7.05[a]	
Δ 5,24 Stigmastadienol%	0.81[a]	0.81[a]	0.56[b]	
Δ 7-Stigmastenol%	0.93[a]	0.71[b]	0.18[c]	≤0.5
Δ 7-Avenasterol%	0.92[a]	0.93[a]	0.67[bc]	
apparent β-Sitosterol %*	93.63[a]	96.08[a]	94.82[a]	≥93

a,b,c: Different superscripts for the same quality parameter mean significant ifferences among varieties (n = 6; $p < 0.01$). nd: not detected.* Apparent β-Sitosterol = β-Sitosterol + Δ5-Avenasterol+Clerosterol+Sitostanol + Δ5,24-Stigmastadienol. Results expressed as percentage of total sterols.

This confirmed the authenticity of the virgin olive oils analyzed in this study (Aparicio and Aparicio-Ruiz, 2000). The highest phytosterol levels found which correspond to β-sitosterol, followed by Δ5-avenasterol, is characteristic of the VOO in the pulp of the olive (Ben Temime et al. 2008).

In this study, β-sitosterol level showed a negative correlation with Δ5-avenasterol level, these two major phytosterols are present in table1 with significant differences ($p < 0.01$).

Ouni et al. (2011) studied the sterolic composition of some Tunisian cultivars and observed that β-sitosterol was the major sterol, with a percentage range from 74.8% to 88.7%, followed by Δ5-avenasterol (4.1 –

19%) and campesterol (1.9 – 2.9%). The studied Tunisian cultivars showed very low amounts of campesterol when compared with those from other studied cultivars, namely the Spanish ones studied by Rivera et al. (2004) and the Portuguese ones studied by Alves et al. (2005).

The minor components cholesterol and Δ7-stigmastenol, which showed no significant differences between studied oils (p < 0.01), were below the established limit of 0.5%, although these are minor components with 24-methylenecholesterol and Δ7-avenasterol in the sterol fraction. 24-methylenecholesterol is an immediate metabolite in the synthesis of campesterol and is characteristic of the oil in the pulp of the olive, but is not found in the oil of the stones (Arafat et al. 2016).

In order to study sterolic compounds might be useful for chemometric analysis, to discriminate between Chemlali and selected new olive varieties VOOs, a principal component analysis (PCA) was performed.

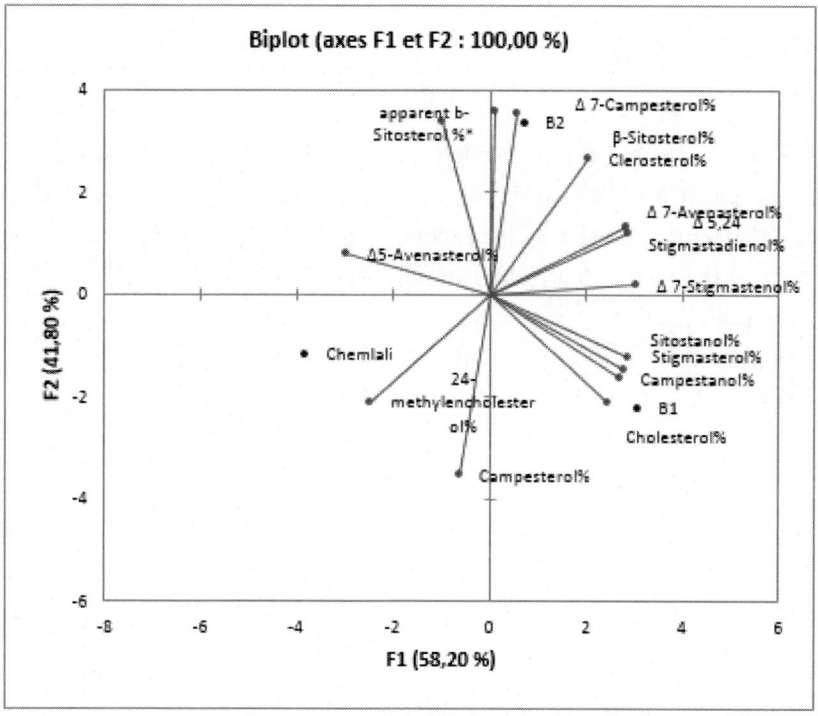

Figure 2. Principal components analysis based on sterolic composition.

The first (F1) and the second (F2) principal components were sufficient to display the structure of the data, since they explained 100% of the total variance.

By examining the scores-plot in the area defined by F1 and F2, VOO samples were separated into three groups, based on their sterolic composition. Group I, was located at the top of the scores-plot, and was formed by B2.

Such group was characterized by the highest levels of, β-sitosterol, Δ7-campesterol and clerosterol. Group II, located on the on the right side of the scores-plot was characterized by having the highest amount of sitostanol compestanol and stigmasterol.

GroupIII, which was located on the left side of the scores-plot, was composed by chemlali and characterized by the highest content of Δ5-Avenasterol (Figure2).

CONCLUSION

The uncontrolled crossing between olive cultivars (autochthonous and foreigner cultivars used as pollinator or pollen acceptor) produce new progenies characterized by an excellent olive oils in terms of sterols, conform to the norm established by EU Regulations. Therefore, the feral olives populations may be a useful tool for improving the Tunisian olive oil composition since it provides new cultivars adapted to climate of Tunisia, resistant to the olive tree diseases and with a good oil quality as compared to Chemlali oil which is characterized by low phenolic compounds contents, low oxidative stability and oleic acid and high linoleic and Palmitic acids. So, these resistant hybrids can be identified and then used for replanting, or as sources for resistance in future breeding programs.

"Reviewed by Pr. Mokhtar Zarrouk, Centre of Biotechnology of Borj Cédria, Tunisia".

REFERENCES

Abumweis, S. S., Barake, R. and Jones, P. J. (2008). Plant sterols/stanols as cholesterol lowering agents: a meta-analysis of randomized controlled trials. *Food Nutr. Res.*, 52, 1 - 17.

Alves, M. R., Cunha, S. C., Amaral, J. S., Pereira, J. A. and Oliveira, M. B. (2005). Classification of PDO olive oils on the basis of their sterol composition by multivariate analysis. *Anal. Chim. Acta*, 549, 166 - 178.

Angiolillo, A., Mencuccini, M. and Baldoni, L. (1999). Olive genetic diversity assessed using amplified fragment length polymorphisms. *Theor. Appl. Genet.*, 98, 411 - 421.

Aparicio, R. and Aparicio-Ruiz, R. (2000). Authentication of vegetable oils by chromatographic techniques. *J. Chromatogr. A*, 881, 93 - 104.

Arafat, S. M., Basuniy, A. M. M., Elsayed, M. E. and Soliman, H. M. (2016). Effect of pedological, cultivar and climatic condition on sterols and quality indices of olive oil. *Scientia Agric.*, 13(1), 23 - 29.

Baccouri, B., Guerfel, M., Zarrouk, W., Taamalli, W., Daoud, D. and Zarrouk, M. (2011). Wild olive (olea europea L.) selection for quality oil production. *J. Food Biochem.*, 35, 161 - 176.

Baccouri, B., Zarrouk, W., Guerfel, M., Baccouri, O., Nouairi, I., Krichene, D., Daoud, D. and Zarrouk, M. (2008). Composition, quality and oxidative stability of virgin olive oils from some selected wild olives (Olea europaea L. subsp. Oleaster). *Grasas Aceites,* 59 (4), 346 - 351.

Boskou, D., Blekas, G. and Tsimidou, M. (2006). Olive oil composition. In D. Boskou (Ed.), Olive oil. *Chemistry and technology,* (pp. 41 - 72). Champaign, IL: AOCS Press. established by EU regulations. *Food Chem.*, 84, 533 - 537.

EU Regulation (2016). Commission Delegated Regulation EU 2016/2095 of 26 September 2016 amending Regulation (EEC) No 2568/91 on the characteristics of olive oil and olive-residue oil and on the relevant methods of analysis. *Off.J. Eur. Union L*, 326, 1 - 6.

Giacalone, R., Giuliano, S., Gulotta, E., Monfreda, M. and Presti, G. (2015). Origin assessment of EV olive oils by esterified sterols analysis. *Food Chem.*, 188, 279 - 285.

Guillaume, C. and Ravetti, L. (2015). Technological and agronomical factors affecting sterols in Australian olive oils. *Riv. Ital. Sostanze Grasse*, 92 (1), 53 - 60.

Mekuria, G. T., Collins, G. G., and Sedgley, M. 1999. Genetic variability between different accessions of some common commercial olive cultivars. *J. Hort. Sci. Biotech.*, 74, 309 - 314.

Morettini, A., Pulselli, A. (1953). In: Lavee, S., 1996. *Biology and physiology of the olive.* In: World Olive Encyclopaedia. International Olive Oil Council, Spain, pp. 61 - 110.

Ouazzani, N., Lumaret, R., Villemur, P. and Di-Giusto, F. (1993). Leaf allozyme variation in cultivated and wild olive tree (Olea europea L.). *J. Hered.*, 84, 34 - 42.

Ouni, Y., Guido, F., Ben Youssef, N., Guerfel, M. and Zarrouk, M. (2011). Sterolic composition and triacylglycerols of Oueslati virgin olive oil: comparison among different geographic areas. *Int. J. Food Sci. Technol.*, 46, 1747 - 1754.

Rivera del Álamo, R. M., Fregapane, G., Aranda, F., Gomez-Alonso, S. and Salvador, M. D. (2004). Sterol and alcohol composition of Cornicabra virgin olive oil: the campesterol content exceeds the upper limit of 4%

Spennemann, D. H. R., Allen, L. R. (2000). Feral olives (Olea europaea) as future woody weeds in Australia: a review. *Aust. J. Exp. Agric.*, 40, 889 - 901.

Stiti, N., Msallem, M., Triki, S. and Cherif, A. (2002). Etude de la fraction insaponifiable de l'huile d'olive de différentes variétés Tunisiennes. *Riv. Ital. Sostanze Grasse*, 79(10), 357 - 363.

In: Sterols: Types, Classification and Structure   ISBN: 978-1-53617-231-7
Editor: Scott Jimenez   © 2020 Nova Science Publishers, Inc.

Chapter 5

STEROLS OF VIRGIN ARGAN OIL: COMPARISON WITH OLIVE OIL

Bechir Baccouri[1] and Imene Rajhi[2]
[1]Centre of Biotechnology of Borj Cédria,
LR15CBBC05 Laboratory of Olive Biotechnology,
B.P. 901, Hammam-Lif, Tunisia
[2]Centre of Biotechnology of Borj Cédria,
Laboratory of Legumes, Hammam-Lif, Tunisia

ABSTRACT

Plant sterols or phytosterols are a family of phytochemicals and common components of legumes, cereals and plant oils, seeds and nuts. They are well known for their cholesterol-lowering effect in humans and have other beneficial effects on health since they inhibit colon cancer development and may prevent some cardiovascular and inflammatory diseases.

The sterolic fraction of argan oil (Argania spinosa L. skeels) and olive oil (Olea europaea L. cv. Chemlali), were investigated and compared. The total phytosterol content ranged from 1700.80mg/kg in chemlali oil to 150.40mg/kg in argan oil. In contrast to chemlali oil in which β-sitosterol is predominant, with 85.8%. The major sterols detected in the argan oils were schottenol and spinasterol. Detection of edible oil adulteration is of utmost importance to ensure product quality and customer protection. Campesterol, a sterol found in argan and Chemlali olive oil, represents less than 0.4% of total sterol content. Interestingly, argan oil contains only traces of campesterol. On the other hand, schottenol and spinasterol were not detected in chemlali oil.

Keywords: Sterol-Argan Oil-olive oil-schottenol-spinasterol

INTRODUCTION

Argan (Argania spinosa L.) and olive (Olea europaea L.), are two crop species appreciated for their fruits and fruit products with high nutritional value for human consumption. Argan oil is traditionally consumed fresh or used for cooking. Because of its high level of unsaturated fatty acids and antioxidants (Gharby et al. 2011), both types of compounds are known to reduce the risks of cardiovascular diseases (El Monfalouti et al. 2010).

Argan is one of the most important oil crops in the world, producing one of the most expensive oils on the market. This tendency is nowadays strongly reinforced by the scientific confirmation of argan oil's potential pharmacological properties (El Monfalouti et al. 2010) and the continuous discovery of anticancer substances in argan oil. In the past few years, the interest given to argan oil has increased in view of its beneficial effects on human health; e.g., antiatherogenic effect (Menendez et al. 2010) cancer chemopreventive effects (Polette et al. 1999) and hypolipemiant and antioxidant properties (El Monfalouti et al. 2010). Recently, we have shown the preventive effect of argan oil against metabolic dysregulation during septic shock in mice model, suggesting new therapeutic potentialities of argan oil in the management of human sepsis (El Babili et al. 2010). However, information on the phytochemical composition of argan oil is still very limited. Similarly to olive oil, previous studies have shown that argan oil contains fatty acids (Drissi et al. 2004), sterols (Hilali et al. 2005) and tocopherols (Drissi et al. 2004).

Olive oil is one of the oldest and most important oils produced by the food industry (Boskou et al. 2006). It is included in the daily diet of the

Mediterranean populations and used for its beneficial effects to human health (Ranalli et al. 2002).

Indeed, olive oil plays a preventive role against many diseases such as dyslipidemia, hypertension, diabetes, obesity and several cancers (Ranalli et al. 2002).

It is well known that olive oil contains high level of oleic acid, which is reported in decreasing the risk of the coronary heart disease; and other minor components with antioxidant and anti-inflammatory properties, including sterols (Lerma-García et al. 2011).

Determination of sterol composition is a well established method and it is also used to detect the adulteration of Argan oils, and it has been recently proposed as a way to classify virgin olive oils according to their fruit variety (Ranalli et al. 2002; Arafat et al. 2016).

The control of sterol fraction is an important issue for vegetable oil genuineness.

Since several studies have demonstrated that each oily fruit has its characteristic sterol profile, its determination could provide abundant information about oil quality (Baccouri et al. 2018) and, therefore, it could be used for authentication purposes. Regarding olive oil, sterol composition has been used for the detection of fraudulent admixtures of olive oil with other cheaper vegetable oils, such as hazelnut, corn, soybean, sunflower and cotton seed oils (Boskou et al. 2006). In addition, this profile has permitted the characterization of olive oils according to their genetic variety and quality grade (virgin, refined, solvent extracted, olive pomace oil, crude olive pomace oil, refined olive–pomace oil, and refined seed oil) (Boskou 2006). However, composition and total sterol contents are strongly influenced by many factors, such as geographical area, crop season agronomic practices, such as irrigation, soil conditions, storage time and temperature (Ben Temime et al. 2006; Guillaume et al. 2012; Boulkroune et al. 2017).

The purpose of this study was to compare the sterolic fractions of Argan and olive oils. The sterolic composition of olive oil (O.europaea L. cv. Chemlali), argan oil (Argania spinosa L. skeels), were analyzed and compared.

METHODS

1. Oil Samples

After harvesting, the olive and argan fruits were immediately transported to the laboratory mill, where the oils were extracted using an Abencor analyzer (MC2 Ingenieriay Sistemas, Sevilla, spain Samples were crushed with a hammer mill and slowly mixed for 30min, centrifuged without addition of warm water, and then transferred into dark glass bottles. All samples were stored at 4°C in darkness in amber glass bottles until analysis.

2. Analytical Determinations

2.1. Determination of Sterols

The qualitative and quantitative sterol contents of the samples were determined according to the European Official Analysis Methods, described in Annexes V and VI of Regulation EEC/2568/91 of the European Union Commission. The determination gave sterols expressed in ppm total sterols only and% individual sterols. The oil sample was saponified with ethanolic potassium hydroxide solution. The unsaponifiable fraction was removed with ethyl ether. The unsaponifiable sterol fraction was separated by chromatography on silica gel plate.

Separation and quantification of the silanised sterol fraction was carried out by capillary gas chromatography, on a Hewlett Packard 6890 chromatograph (Hewlett-Packard, Agilent, CA) with autosampler and flame ionisation detector (FID) using an HP-5MS capillary column (30m×0.25mm×0.25μm). The working conditions of the chromatograph were: injector 300°C, isothermal column 260°C, and detector 325°C. The injected quantity was 0.2μl, at a flow rate of 1.1 ml/min, using helium as carrier gas. The amount of oil used was 5g.

Chromatographic run time 40mn was. IS amount was 2 – 3μl in each sample. Quantification was made by addition of an internal standard (α-cholestanol, 0.2%) and apparent β-sitosterol was calculated as the sum of β-sitosterol, Δ 5,23-stigmastadienol, clerosterol, sitostanol and Δ5,24-stigmastadienol. Sterols peak identification was carried out according to the reference method.

2.2. Statistical Analysis

The results are reported as the mean values. Data were compared on the basis of standard deviation of the mean values. In addition, Duncan's multiple range tests were used to determine significant differences among data. All collected data were submitted to hierarchical clustering analysis. The associations obtained based on the similarity in Euclidian distances using the Statistica 5.0 package. Principal components analysis is performed with XLSTAT software, Version 2014.

RESULTS AND DISCUSSION

Sterols, which comprise a major portion of the unsaponifiable matter, are found in almost all fats and oils, and they are also characteristic of the genuineness of vegetable oils. The composition of the steroidal fraction of the olive oil is a very useful parameter for detecting adulterations or to check authenticity, since it can be considered as its real fingerprint.

The total phytosterol content ranged from 1700mg/Kg in olive oil to 1600.40mg/kg in argan oil (Table 1). Results show that the main sterols found in studied olive oil were β sitosterol, Δ5-avenasterol and campesterol. β-sitosterol was the main compound found in olive oil, with 85.89% (Table 1). Our results are consistent with those of Baccouri et al. (2018), who found that β-sitosterol is the main phytosterol in Tunisian wild olive oil, with a range of 71.01 – 87.18%, relative to the genotype.

β Sitosterol is the most abundant sterol in olive oil and has a recognised effect in lowering cholesterol levels by opposing its absorption in the intestinal tract. The health aspects of β sitosterol, Major sterol of olive oil, have also been reported in several studies (Matos et al. 2007; Lukić et al. 2013).

Table 1. Principal components analysis based on sterolic composition

	Olive oil	Argan oil	Norm
Cholesterol%	0.11a	0.18a	≤0.5
24-methylencholesterol%	0.18dc	0.17b	
Campesterol%	3.60a	0.4c	≤4
Campestanol%	0.05c	3.6b	
Stigmasterol%	0.39ef	nd	Campesterol
Clerosterol%	0.94b	0.6a	
β-Sitosterol%	85.89bc	3.5f	
Sitostanol%	0.38d	0.58e	
Δ5-Avenasterol%	7.05dc	0.5b	
Δ 5,24 Stigmastadienol%	0.56d	nd	
Δ 7-Stigmastenol%	0.18g	4.5	≤0.5
Δ 7-Avenasterol%	0.67bc	nd	
the apparent βSitosterol	94.82	1.68	
Spinasterol	nd	31.5	
Schotenol	nd	43.3	
δ8,22-Stigmastadiene-3β-ol	nd	0.45	

a,b,c: Different superscripts for the same quality parameter mean significant ifferences among varieties (n = 6; p < 0.01). nd: not detected.* Apparent β-Sitosterol = β-Sitosterol + Δ5-Avenasterol + Clerosterol + Sitostanol + Δ5,24-Stigmastadienol. Results expressed as percentage of total sterols.

They refer mainly to the reduction of cholesterol levels and the prevention of many diseases and various types of cancer (colon, prostate, and breast).

On the other hand, the apparent βSitosterol, expressed by the sum of the contents of β sitosterol and other four sterols formed by the degradation of β sitosterol (sitostanol, Δ5,24-stigmastadienol, clerosterol and Δ5-avenasterol), was higher in olive oil than that of argan oil. Moreover, schottenol and spinasterol are the main sterols found in argan oil (43.3 and 31.5%, respectively). This is in good agreement with the results reported by Matthäus et al. (2011). Schottenol and spinasterol were not detected in olive oil (Table 1). Schottenol and spinasterol might have health-promoting properties by modulating the metabolism of cholesterol (Matthäus et al. 2011).

The composition of the sterol fraction of olive oil is a very useful parameter for detecting adulterations or to check authenticity, since it can be considered as a fingerprint (Marini et al. 2004. Relative to the authenticity indices established by current legislation (EEC 1991), the campesterol content was remarkably high. These compounds showed changes in the studied oils according to the species. The campesterol content does not exceed the upper limit established by the EU Regulation (4%). The percentages of stigmasterol in all tested samples were lower than those of campesterol, this result showed that all the oil samples came from healthy fruit not obtained by systems of forcing (Berger et al. 2004). The highest campestanol content was found in argan oil (3.6). Hence, The stigmasterol, Δ 7-Avenasterol and Δ 5,24 Stigmastadienol compounds were not detected in argan oil.

These findings are in good agreement with those of other authors (El Kharrassi et al. 2002). They reported that the total phytosterols content ranged from 203.80mg/100g in olive oil to 2131.40mg/100g in the cactus cladode essential oil (CCEO). Argan oil have an intermediate content (160mg/100g. Therefore, β-sitosterol was the main phytosterol found in olive oil and cactus pear seed oil. The argan oil was rich in spinasterol and schottenol while the CCEO contains high level of sitostanol.

In order to study how the studied parameters are useful in chemometric analysis to discriminate between studied oils, a principal component analysis (PCA) was performed. The first and the second principal components were sufficient to display the data structure, since they explained 100% of the total variance. By examining the scores-plot in the area defined by the first and the second principal components, the samples were separated into two groups based on the studied parameters. Group I, which is located on the left side of the scores-plot, is composed of argan oil. Group II is located on the right side of the scores-plot and include olive oil. Since the samples were well described by the scores-plot, the loading plot (Figure 2) was analysed in order to show which variables influenced the group separation. Group I was characterised by the highest content of

Schottenol and spinasterol. Group II was characterised by a high level of the apparent βSitosterol compounds (Figure 2).

Four sterols have been isolated from argan oil (Farines et al. 1984), spinasterol, schottenol, (3β,22E, 24S)-stigmasta-5,22- dien-3- ol, and (3β,24Z)-stigmasta-7,24–28-dien-3-ol (Figure 1).

The total content of sterols in the unsaponifiable fraction of argan oil is approximately 20%. Farines and colleagues (1981), Charrouf and Guillaume (1999), Khallouki (2003), Khallouki.

and colleagues (2003) report that argan oil contains spinasterol (40%) and its dihydrospinasterol (schottenol, 48%) as major sterols, respectively, together with Δ7-avenasterol and stigmasta-8,22-diene-3-β-ol in lower concentrations. Spinasterol and schottenol are rarely found in vegetable oils. Spinasterol has been described as the characteristic phytosterol of the sapotaceae family (Gunasekera et al. 1977).

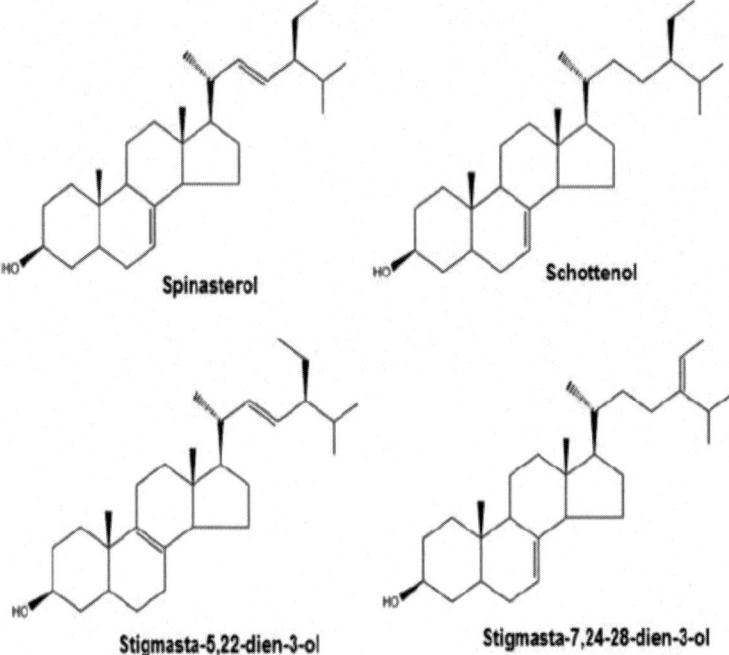

Figure 1. Sterols present in argan oil (Charrouf and Guillaume 2002).

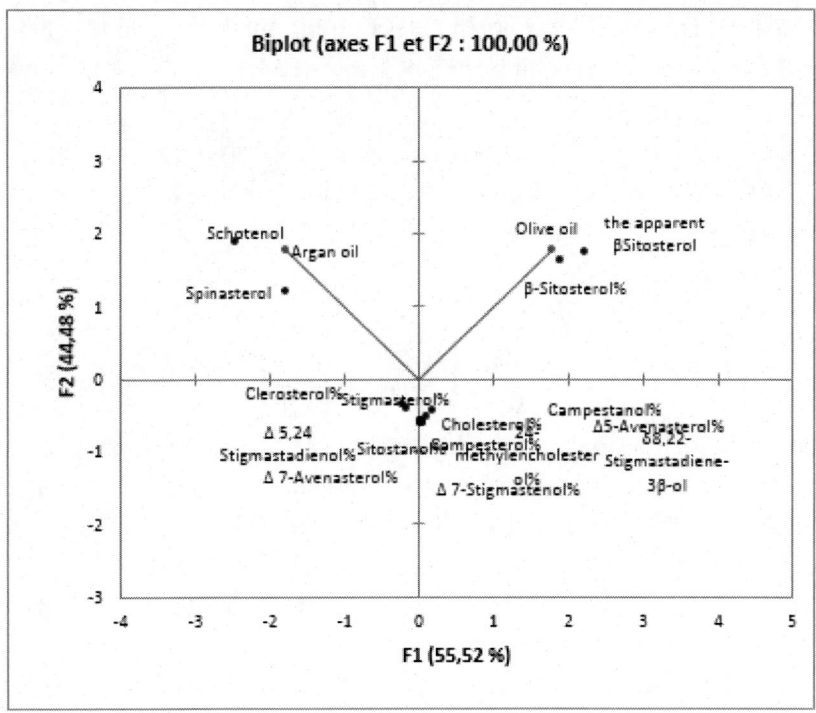

Figure 2. Principal components analysis based on sterolic composition.

Contrary to the composition of fatty acids, the phytosterol composition is very different from that of sesame and peanut oils in which β-sitoterol dominates.

CONCLUSION

The analysis has shown that there were qualitative and quantitative differences between the sterolic profiles of the tested olive oil compared with that of argan. The results indicate that genetic factors are one of the most important aspects of the sterolic composition of oil. From this analysis, we can conclude that the studied oils have an excellent composition in terms of sterols. In addition, they fully conform to EU regulations (EEC 2003). The argan oil is rich in spinasterol and schottenol while the olive oil contains high level of β Sitosterol.

"Reviewed by Pr. Mokhtar Zarrouk, Centre of Biotechnology of Borj Cédria, Tunisia".

REFERENCES

Arafat, S. M., Basuniy, A. M. M., Elsayed, M. E. and Soliman, H. M. (2016). Effect of pedological, cultivar and climatic condition on sterols and quality indices of olive oil. *Scientia Agric.,* 13(1), 23 - 29.

Baccouri, B., Manaia, H., Casas, J. S., Osorio, E. and Zarrouk, M. (2018). Tunisian wild olive (Olea europaea L. subsp. oleaster) oils: Sterolic and triterpenic dialcohol compounds. *Ind. Crops Prod.,* 120, 11 - 15.

Ben Temime, S., Manai, H., Methenni, K., Baccouri, B., Abaza, L., Daoud, D., Sanchez-Casas, J., Osorio-Bueno, E. and Zarrouk, M. (2008). Sterolic composition of Chetoui virgin olive oil: influence of geographical origin. *Food Chem.,* 110, 368 - 374.

Boskou, D. (1996). Olive oil chemistry and technology. AOCS press, Champaign, pp. 12 - 52.

Boulkroune, H., Lazzez, A., Guissous, M., Bellik, Y. Smaoui, S., Grati Kamoun N. and Madani, T. (2017). Characterization of sterolic and alcoholic fractions of some Algerian olive oils according to the variety and ripening stage. OCL-Ol Corps Gras Li. (DOI: 10.1051/ocl/2017026).

Drissi, A., Girona, J., Cherki, M., Godas, G., Derouiche, A., El Messal, M., Saile, R., Kettani, A., Sola, R., Masana, L., Adlouni, A. (2004). Evidence of hypolipemiant and antioxidative properties of argan oil derived from the argan tree (Argania spinosa). *Clin. Nutr.,* 23:1159 - 1166.

Drissi, A., Gironab, J., Cherki, M., God`as, G., Derouiche, A., El Messal, M., Saile, R., Kettania, A., Sol`a, R., Masanab, L. and Adlouni, A.

(2004). Evidence of hypolipemiant and antioxidant properties of argan oil derived from the argan tree (*Argania spinosa*). *Clin. Nutr.*, 23:1159 - 1166.

El Babili, F., Bouajila, J., Fouraste, I., Valentin, A., Mauret, S. and Moulis, C. (2010). Chemical study, antimalarial and antioxidant activities, and cytotoxicity to human breast cancer cells (MCF7) of *Argania spinosa*. *Phytomedicine*, 17:157 - 160.

El Monfalouti, H., Guillaume, D., Denhez, C. and Charrouf, Z. (2010). Therapeutic potential of argan oil: A review. *J. Pharm. Pharmacol.*, 62:1669 - 1675.

EU Regulation (2016). Commission Delegated Regulation EU 2016/2095 of 26 September 2016 amending Regulation (EEC) No 2568/91 on the characteristics of olive oil and olive-residue oil and on the relevant methods of analysis. *Off. J. Eur. Union L*, 326, 1 - 6.

Gharby, S., Harhar, H., Guillaume, D., Haddadb, A., Matth¨ausd, B. and Charroufa, Z. (2011). Oxidative stability of edible argan oil: A two year study. *LWT-Food Sci. Technol.*, 44:1 - 8.

Guillaume, C., Ravetti, L., Ray, D. L. and Johnson, J. (2012). Technological factors affecting sterols in Australian olive oils. *J. Am. Oil Chem. Soc.*, 89, 29 - 39.

Hilali, M., Charrouf, Z., El Aziz Soulhi, A., Hachimi, L., Guillaume, D. (2005). Influence of origin and extraction method on argan oil physico-chemical characteristics and composition. *J. Agric. Food Chem.*, 53: 2081 - 2087.

Lerma-García, M. J., Simó-Alfonso, E. F., Méndez, A., Lliberia, J. L. and Herrero-Martínez, J. M. (2011). Classification of extra virgin olive oils according to their genetic variety using linear discriminant analysis of sterol profiles established by ultra-performance liquid chromatography with mass spectrometry detection. *Food Res. Int.*, 44, 103 - 108.

Lukić, M., Lukić, I., Krapac, M., Sladonja, B. and Piližota, V. (2013). Sterols and triterpene diols in olive oil as indicators of variety and degree of ripening. *Food Chem.*, 136, 251 - 258.

Matos, L. C., Cunha, S. C., Amaral, J. S., Pereira, J. A., Andrade, P. B., Seabra, R. M. and Oliveira, B. P. P. (2007). Chemometric characterization of three varietal olive oils (Cvs. Cobrançosa, Madural and Verdeal Transmontana) extracted from olives with different maturation indices. *Food Chem.*, 102, 406 - 414.

Matthäus, B. and Özcan, M. M. (2011). Determination of fatty acid, tocopherol, sterol contents and 1,2- and 1,3-diacylglycerols in four different virgin olive oil. *J. Food Process Technol.*, 2, 1 - 4.

Menendez, J. A., Vellon, L., Colomer, R., Lupu, R. (2005). Oleic acid, the main monounsaturated fatty acid of olive oil, suppresses her-2/neu (erb b-2) expression and synergistically enhances the growth inhibitory effects of trastuzumab (herceptinTM) in breast cancer cells with her-2/neu oncogene amplification. *Ann. Oncol.*, 16:359 - 371.

Polette, M., Huet, E., Birembaut, P., Maquart, F. X., Hornebeck, W., Emonard, H. (1999). Influence of oleic acid on the expression, activation and activity of gelatinase as produced by oncogenetransformed human bronchial epithelial cells. *Int. J. Cancer*, 80:751 - 755.

Ranalli, A., Pollastri, L., Contento, S., Di Loreto, G., Iannucci, E., Lucera, L. and Russi, F. (2002). Sterol and alcohol components of seeds, pulp, and whole olive fruit oils. Their use to characterize olive fruit variety by multivariates. *J. Sci. Food Agric.*, 82, 854 - 859.

In: Sterols: Types, Classification and Structure ISBN: 978-1-53617-231-7
Editor: Scott Jimenez   © 2020 Nova Science Publishers, Inc.

Chapter 6

USE OF PHYTOSTEROLS AS A TOOL FOR THE AUTHENTICITY ASSESSMENT OF VIRGIN OLIVE OIL: PROTECTION OF THE OLIVE OIL MARKET

Imen Oueslati[*], *Hedia Manai-Djebali* and *Ridha Mhamdi*

Centre of Biotechnology of Borj-Cédria,
LR15CBBC05 Laboratory of Olive Biotechnology,
Hammam-Lif, Tunisia

ABSTRACT

Olive oils are distinct from other vegetable oils because they may be consumed without extensive refining. Virgin olive oils attract a higher price than refined olive oils because of their pleasant, rather delicate flavour and aroma and limited production volume. As a result, olive oils are subject to two types of deliberate adulteration. The first is the blending of virgin olive oils with olive oils of lower grade (e.g., refined

---

[*] Corresponding Author: imen.oueslati@fst.rnu.tn.

olive oil or olive-pomace oil). The second is the less subtle mixing of olive oil with liquid vegetable oils. As a consequence of such adulteration, the International Olive Oil Council (IOOC) and the Codex Alimentarius Commission (CAC) have produced standards for virgin and refined olive and olive-pomace oils and certain blends of these products. Along with the profile of fatty acids, triglycerides and tocopherols, the sterol profile is an important parameter to assess the identity and authenticity of fats and oils; it is widely accepted as one of the most important markers for the detection of adulterated olive oils. During the analysis of the sterolic profiles, the most frequent deviations observed in the samples of extra virgin olive oil were in: campesterol, Δ7-stigmastenol, apparent β-sitosterol, total sterols, erythrodiol and uvaol. In the case of the lampante samples, deviations were noted in the following criteria: Δ7-stigmastenol, apparent β-sitosterol, total sterols, erythrodiol + uvaol. Only two parameters deviated from the official limits in the olive pomace oils tested: Δ7-stigmastenol and apparent β-sitosterol. Campesterol and Δ7-stigmastenol were the parameters for which the most samples exhibited deviations. These results are very important in fighting fraud and ensuring that the olive oil that consumers buy for its health or sensory properties has not been mixed with other, cheaper vegetable oils. The purpose of this paper, however, is to show how the standards of the European Union legislation for fats and oils can help to verify olive oil authenticity using the sterolic profile. This review proposes possible solutions to safeguard the consumer and protect the olive oil market.

Keywords: olive oil, sterols, erythrodiol and uvaol, authenticity, adulteration, traceability

INTRODUCTION

Since the turn of the century, vegetable oils have supplanted lard and beef tallow as the major source of dietary fat. The change in consumer preferences is mostly due to consumers demanding food products that combine a pleasant flavor with nutritional benefits.

Olive oil is recognized worldwide for its nutritional value, health benefits, desirable organoleptic properties and rich concentration of bioactive constituents (Elloumi, et al., 2012). Olive oil is considered an economically important crop, especially for producing countries in the Mediterranean area. The olive oil varieties consumed in the Mediterranean basin retain virtually all their natural nutritional properties because they are usually obtained from their respective plant sources through either physical crushing or pressing.

Due to the heterogeneity of the physical product and its quality categorization, "extra virgin olive oil (EVOO)" is the grade of olive oil most highly valued for its superior taste and reported health benefits; it is the most expensive olive oil to produce. EVOO is, also, considered a price premium product compared to other vegetable oils.

However, the discrepancy in pricing between EVOO and other commodity oils has rendered this product a primary target for fraudulent activities, namely economic adulteration and deliberate mislabeling (Srigley, et al., 2016). Some reports estimate that adulteration of EVOO with hazelnut oil in the European Union (EU) is a multimillion euro operation annually (Azadmard-Damirchi, et al., 2005). Opportunists have reportedly also adulterated olive oil with canola (rapeseed), soy, mustard, other seed oils, and refined olive oil (Frankel, 2010). This procedure, however, is also harmful since consumers buy olive oil for its health benefits and are surprised to receive oil that does not provide them. The paradigm was the toxic oil syndrome resulting from consumption of olive oil spiked with aniline-denatured rapeseed oil that affected more than 20 000 people.

Adulteration of virgin olive oil lead to economical losses, disloyal competition among producers, and break in consumer trust. The authenticity of the VOO become an important subject from both a commercial and a health perspective. Authenticity (Lehman, 2019) covers many aspects, including adulteration, mislabeling, characterization and misleading origin.

Some cultivars of olive oil are recognized as being of higher quality because they derive from well defined geographical areas, command better prices, and generally are legally protected. Indeed, the aim of protected designations of origin (PDO) is to add value to certain specific high-quality products from a particular origin. The definition of traceability according to the European Council Regulation (EEC, 178/2002) is the ability to identify and trace a product or a batch of products at all stages of production and marketing. Traceability is important for commercial reasons and plays a considerable role in the assurance of public health.

Over the last 10 years, research and technology have experienced great progress in the fight against forgery of EVOOs. Nevertheless, the knowledge of fraudsters has also followed this trend, which allowed the introduction of more sophisticated frauds that require the use of novel approaches for detection. Biochemical techniques have been employed for the classification and authenticity of olive oils using a high number of variables such as triacylglycerides (TAGs) composition, phenolic fraction, DNA markers and sterolic profile (Busconi, et al., 2003).

The purpose of this paper is to present the usefulness of the sterolic profile of the vegetable oil to detect the adulteration of the virgin olive oil. Several sophisticat equipements were used to detect the sterolic markers of the adulteration and the authentication of the VOO. This study illustrates the importance of sterolic molecular markers to study the composition and the traceability of olive oils.

PHYTOSTEROLS

Plant sterols, also called phytosterols, comprise a major proportion of the unsaponifiables in vegetable oils. They are biosynthetically derived from squalene and form a group of triterpenes (Goodwin, 1980). They are important components of plant cells in controlling membrane fluidity and permeability, although some have a specific function in signal transduction events and membrane-bound enzyme activity (Piironen, et al., 2000). The phytosterol content of vegetable oils can differ with cultivar, crop year, ripening, storage time, extraction methods, etc. Commercially, phytosterols are isolated from vegetable oils, such as soybean oil, rapeseed (canola) oil, sunflower oil or corn oil, or from so-called "tall oil", a by-product of the manufacture of wood pulp. Phytosterols are derivatives of a tetracyclic perhydro-cyclopentano-phenanthrene ring system. They are structurally related to cholesterol but differ from cholesterol in the structure of the side chain. They consist of a steroid skeleton with a hydroxyl group attached to the C-3 atom of the A-ring and an aliphatic side chain attached to the C-17 atom of the D-ring. Sterols have a double bond, typically between C-5 and C-6 of the sterol moiety. Phytosterols are classified into three classes based on the presence or absence of methyl groups at the C-4 position in the A ring: 4-desmethylsterols (without methyl group), 4-monomethylsterols (one methyl group) and 4,40-dimethylsterols (triterpene alcohols, two methyl groups). The structural formulae of sitosterol (4-desmethylsterol), citrostadienol (4-monomethylsterol) and 24-methylenecycloartanol (4,40-dimethylsterol) are shown in Figure 1. Methylsterols (4-monomethyl- and 4,40-dimethylsterols) are synthesized at an early stage in the biosynthetic pathway and are precursors of 4-desmethylsterols. Phytosterols can occur in free and esterified forms, i.e., as fatty acid esters, steryl glycosides or acylated steryl glycosides. In free-form, the hydroxyl group at the C-3 in the A ring is underivatised, whereas, in esterified form, the hydroxyl group is covalently bound to other constituents.

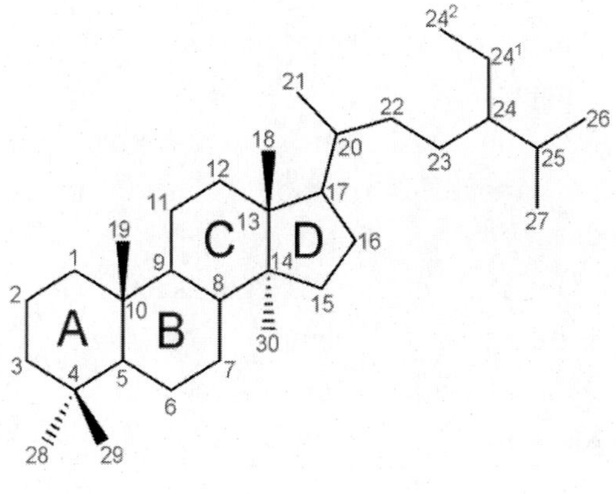

Figure 1. Steroid skeleton.

The conventional method for phytosterol analysis is saponification of the oil sample followed by extraction of the unsaponifiables with an organic solvent. This method can quantify the total amount of phytosterols without providing any information on the distribution of free or esterified forms.

On the other hand, a separate determination of phytosterols in free and esterified forms provides detailed information as to their distribution and stability. Separation of free and esterified phytosterols can be done with Thin-Layer Chromatography (TLC), High-performance Liquid Chromatography (HPLC) or Solid Phase Extraction (SPE). The latter method is simple and rapid; for example, using SPE (1 g silica), esterified sterols (less polar) and free sterols (more polar) can be eluted with hexane and a mixture of hexane and diethyl ether, respectively (Azadmard Damirchi and Dutta, 2007).

STEROLIC COMPOSITION OF VIRGIN OLIVE OIL

The unsaponifiable fraction of EVOO, which consists of sterols (desmethylsterols, monomethylsterols, dimethylsterols), hydrocarbons, aliphatic alcohols, and tocopherols, serves as a valuable tool, or fingerprint, for evaluating authenticity (Tena et al., 2015). However, the desmethylsterol composition of EVOO has also been shown to vary by cultivar, growing region, and ripeness of the olive fruits (Tena, et al., 2015), thus complicating the role of desmethylsterols for evaluating authenticity.

Total sterol content was calculated as the sum of the analyzed contents of individual desmethylsterols, namely cholesterol, campesterol, 24-methylene cholesterol, campestanol, brassicasterol, stigmasterol, $\Delta 7$-campesterol, clerosterol (+$\Delta 5,23$-stigmastadienol), β-sitosterol, sitostanol, $\Delta 5$-avenasterol, $\Delta 5,24$-stigmastadienol, $\Delta 7$-avenasterol, and $\Delta 7$-stigmastenol (IOOC, 2015). Contents of individual desmethylsterols were expressed as a percentage of the total sterol content. The content of apparent β-sitosterol was calculated as the sum of the contents of clerosterol, $\Delta 5,23$-stigmastadienol, β-sitosterol, $\Delta 5,24$-stigmastadienol, sitostanol, and $\Delta 5$-avenasterol (IOOC, 2015). The content of triterpene dialcohols, which are most abundant in the epicarp (the outermost portion) of olives, was calculated as the sum of the contents of erythrodiol and uvaol, taken as a percentage of the sum of the total sterol content plus the contents of erythrodiol and uvaol (IOOC, 2015).

STEROLIC MARKERS FOR EVALUATING AUTHENTICITY

The assessment of authenticity was based on a subset of purity criteria specified in the IOOC (2015) standards for grades of olive oil and olive-pomace oil. Eight parameters were used to assess purity, namely total sterol content ($\geq$ 1000 mg/kg) and the concentrations (as % of total sterols) of brassicasterol ($\leq$ 0.1%), campesterol ($\leq$ 4.0%), cholesterol ($\leq$ 0.5%), Δ7-stigmastenol ($\leq$ 0.5%), stigmasterol $\leq$ campesterol, apparent β-sitosterol $\geq$ 93.0, and erythrodiol plus uvaol ($\leq$ 4.5%).

Oils failing to meet purity criteria for total sterol content indicated possible adulteration with desterilized oil (i.e., edible oils in which the sterol fraction is removed to pass undetected for adulteration) (Aparicio, et al., 2013). Oils failing to meet purity criteria for desmethylsterol composition indicated that the concentration of erythrodiol plus uvaol serves as an indicator of possible adulteration with solvent-extracted oils (Grob, et al., 1990). These triterpene dialcohols have also been shown to vary with cultivar, growing region, and maturation stage of the olive fruit (Lukic, et al., 2013). The solvent extraction leads to high amounts of triterpenic dialcohol in refined olive oil, whereas, they are absent or present in low levels in EVOO. Aparicio et al. (2013) recommended that additional confirmatory tests, such as wax or aliphatic alcohol content, be performed to further investigate authenticity when high concentrations of triterpene dialcohols are observed. Moreover, Grob et al. (1999) proposed that the content of wax esters was a more suitable marker for detecting adulteration with solvent-extracted oil.

Elevated levels of certain sterolic compounds in olive oil are indicative of adulteration by seed or refined oil. For example, cholesterol occurs only in very small amounts in plant oils, and olive oil adulterated with soybean, grapeseed, or sunflower oil can be detected on the basis of the concentrations of campesterol and stigmasterol (Kamm, et al., 2001).

Brassicasterol has been used to detect olive oil adulteration with rapeseed oil (Kamm, et al., 2001). As such, the ability to accurately and efficiently determine the sterol composition of olive oil is useful in detecting a variety of potential adulterants.

Campesterol is considered a major diagnostic analyte for detecting adulteration with commodity oil due to the abundant concentration of this desmethylsterol in many commodity oils (Kamm, et al., 2001, Than, et al., 2006). High concentrations of campesterol have been useful for detecting adulteration of Chemlali EVOO with 10% soybean oil and 10% corn oil (Bello, 1992). Likewise, in a study of virgin olive oils from Jordan, high concentrations of campesterol were important for detecting intentional adulteration with 10% cottonseed, soybean, and sunflower oils, and also corn oil at the 5% level (Amelio, 1992). Moreover, the increased concentration of campesterol, among others, was likely the primary cause for the concurrent reduction in apparent β-sitosterol concentration.

4,4-Dimethylsterols have been used to detect virgin olive oil adulteration with pomace olive oil at levels as low as 5% (Ntsourankoua, et al., 1994).

The concentration of Δ7-stigmastenol was also found to be indicative of adulteration with soybean, canola, and safflower oils. Concentrations of Δ7-stigmastenol are especially high in sunflower (180–500 mg/kg) and safflower oils (325–540 mg/kg) oils, whereas in EVOO the concentration falls below 10 mg/kg (Benitez-Sanchez, et al., 2003). Jabeur et al. (2014) found that the concentration of Δ7-stigmastenol exceeded the threshold value of 0.5% when as little as 1% safflower oil was added to Chemlali EVOO. Differences in findings between the present study and those of Jabeur et al. (2014) highlight the importance of the desmethylsterol composition of the EVOO base oil, to which adulterant is added, for masking adulteration.

4-Desmethylsterols have been used to detect olive oil adulteration with vegetable oils at levels as low as 5% (Bohačenko and Kopicova, 2001). Vichi et al. (2001) used a combination of data from free Δ7-sterols (Δ7-stigmastenol and Δ7-avenasterol) and Δ-Equivalent Carbon Number 42 (ΔECN42) to detect adulteration at a level of 10%. Data obtained from analysis of free Δ7-sterols or ΔECN42 were not sufficient alone for this purpose. A spectrofluorimetric method combined with multivariate analysis has been used to assess the authenticity of olive oil in admixtures with hazelnut oil (Sayago, et al., 2004). Stepwise linear discriminant analysis applied to each admixture showed that this method can be used to detect hazelnut oil at levels higher than 5%.

Lanzon and Albi (1989) proposed the use of the determination of stigmastadienes to detect oils that had been refined. Cert et al. (1994) identified stigmasta-3,5-diene, as the main compound of interest, it is used to detect oils that had been refined. Stigmasta-3,5-diene is synthetized through the deodorization refining process. The formation of stigmasta-3,5-diene from β-sitosterol and therefore stigmasta-3,5-diene contents are strongly dependent of the temperature used during the physical refining (León-Camacho, et al., 2004). Stigmasta-3,5-diene has been considered potentially remarkable on the detection of refined olive oil (ROO) in VOO (León-Camacho, et al., 2004). However, stigmastadienes could be only seen as reliable indicators of olive oil adulteration when their concentrations ranged between 0.01 and 4 mg/kg (Al-Ismail, et al., 2010) limiting their use on the detection of refined olive oil and olive oil blends.

PARAMETERS AFFECTING THE STEROLIC PROFILE OF OLIVE OIL

The present community provisions make reference to the determination of some chemical indices to check the authenticity of olive oils, but these indices, separately considered, are unable to supply complete information on the quality of the product. The variation of the chemical composition of the olive oil depend on the geographical origin, maturity stage, extraction method, olive fruit variety (cultivar), etc.

The sterol fraction can be used to gauge maturation, quality, type, and geographic origin of cultivar used to produce olive oil (Fernández-Cuesta, et al., 2013). Stigmasterol and Δ-5-avenasterol have been shown to increase as the olive fruits achieve an advanced maturity stage which coincides with the maximum oil yield of the olive fruit, and thus provide a potential gauge for determining optimal harvest time (Kamm, et al., 2001). The concentration of Δ5-avenasterol is important for olive oil quality, as the presence of an ethylidene group in this sterol confers exceptional antioxidant properties at high temperatures (Rossell, 2001). Total sterol content and profile in olive oil largely depends upon the cultivar, environmental conditions, stage of maturity of olives, and extraction method (Aparicio and Luna, 2002).

Several researches conducted on oils extracted from olives at different maturity stages was characterized by a decrease in the proportion of β-sitosterol and an increase in the Δ5-avenasterol (Vekiari, et al., 2010). A hypothesis to explain the changes in β-sitosterol and Δ5-avenasterol concentrations was postulated by Aparicio and Luna (2002), who suggested that the variability in sterol profile might be the results of the reduction in the proportion of stone in olive weight during ripening, as the oil from stones has a high concentration of β-sitosterol but a low concentration of Δ5-avenasterol (Aparicio and Luna, 2002).

ADULTERATION OF OLIVE OIL WITH VEGETABLE OILS

One of the most common adulteration practices consists of blending EVOO with refined olive oil (Fragaki, et al., 2005; Frankel, 2010) which is obtained usually from VOO mechanically extracted from damaged olive fruits or from olives stored in unsuitable conditions and using refining methods that does not lead to alterations in the initial glyceridic structure. It has a free acidity, expressed as oleic acid, of not more than 0.3 g per 100 g (IOOC, 2011).

Plant sterols or phytosterols make up the main part of the unsaponifiable fraction of olive oil. The most abundant olive oil sterol is β-sitosterol, followed by Δ5-avenasterol. Campesterol and stigmasterol are present in lower concentrations (Harwood and Aparicio, 2000). Regarding the tracking of commercial fraud, the sterol fraction has many applications, especially where the contamination of some vegetable oils with other cheaper ones is concerned. Positional isomers of the double bond in the sterol ring have been detected and can be used as fraud tracers in VOO (Aparicio and Aparicio-Ruiz, 2000).

Contamination of sunflower oil higher than 5% could be detected by the increase of campesterol (7–13% for sunflower oil and 8–10% for high oleic sunflower oil), stigmasterol (8–11% for sunflower oil), and Δ7-stigmastenol (7–13% for sunflower oil and 14–22% for high oleic sunflower oil vs. ≤ 0.5% for olive oil) concentrations. Additions of around 2% rapeseed oil were detected by determining the content of brassicasterol (≤ 0.1% of total sterols for olive oil vs. 12–13% for rapeseed oil and 5–13% for canola oil).

Adulteration with 10% or higher of grapeseed oil would increase the concentrations of campesterol (9–14% for grapeseed oil) and stigmasterol (9–17% for grapeseed oil) but an admixture of around 5% would be at the limits of detection. An alternative to these kinds of sterols is offered by steryl esters, which are complex mixtures that differ in composition from that expected by a random esterification of total fatty acids and sterols in the oil.

Detecting the adulteration of olive oil with hazelnut oil has proven to be a major challenge due to compositional similarities between the two oils (e.g., triacylglycerols, fatty acids, desmethylsterols) (Calvano, et al., 2012). Calvano et al. (2012) noted difficulties in detecting the presence of hazelnut oil in olive oil at adulteration levels less than 20%. Thus, we were surprised to find that one of the three hazelnut oils were detected at the 10% level of adulteration, suggesting that this oil, labeled as an organic virgin hazelnut oil, may have itself been mislabeled. In contrast, Mariani et al. (2006) showed that concentrations of esterified campesterol, Δ7-stigmastenol, and Δ7-avenasterol are useful for detecting adulteration with crude hazelnut at concentrations below the 10% level. Azadmard-Damirchi et al. (2005) proposed that the dimethylsterol fraction may be a more suitable marker for adulteration with crude hazelnut oil due to its contents of lupeol and an unidentified compound with lupine skeleton that are not observed in neat virgin olive oil.

The addition of 10% soybean oil was detected by determining the concentration of campesterol (15–24% for soybean oil vs. 4% for olive oil) and stigmasterol (16–19% for soybean oil vs. less than campesterol for olive oil).

According to Jabeur et al. (2014), the addition of 2% sunflower oil to EVOO was detected by an increased concentration of Δ7-stigmastenol, whereas in Al-Ismail et al. (2010), the addition of 5% corn oil to EVOO was found to increase the concentration of campesterol beyond the threshold value for detection. In both of those studies, adulteration with soybean oil at the 10% level was the lowest concentration at which the presence of adulterant could be detected.

The adulteration of the olive oil with bleached oil can be detected by the presence of the sterolic products of degradation. Bleaching of oils results in a partial dehydration of sterols, and the emission of the new products, Δ3,5-steradienes, which are highly characteristic indicator compounds for industrially refined oils (Cert and Lanzón, 1994), By using "hard" refining, i.e., a high amount of bleaching earth and high temperatures, sterols may completely be removed (desterolized oils), though degradation products remain in the oil (Essid, et al., 2016).

According to the International Olive Oil Council (IOOC), the total sterol content for edible VOO must be at least 1000 mg/kg (0.1%, m/m), and the erythrodiol and uvaol contents must not be greater than 4.5% of the total sterol content (Gunstone, 2011).

METHODS OF STEROLIC ANALYSIS USED FOR THE DETECTION OF OLIVE OIL ADULTERATION

The detection of adulteration poses significant challenges for EVOO due to the diversity of cultivars grown around the world and the limitations of existing official methods for detecting adulteration. During the last decades, monitoring the authenticity of the olive oil is carried out using advanced and sophisticate technology that provide data about their qualitative and quantitative composition (Tena, et al., 2015). The advances in knowledge and technology have certainly led to greater success in the fight against adulteration over the years.

The more commonly used methods for the sterol compounds in olive oil usually require isolation and several procedures of separation, identification and quantification. The conventional method for quantifying sterols and triterpenic alcohols involves capillary Gas Chromatography (GC) with Flame-Ionization Detection (FID) of the fraction isolated by TLC.

To improve the efficiency of the analysis of the sterol composition of oils, several methods were carried out using the HPLC−MS, and/or time-consuming methods, such as TLC cleanup. The unsaponifiable fraction can be separated with a diatomaceous earth column, and the sterol and triterpenic dialcohols were isolated with a base-activated SPE cartridge cleanup protocol. Quantitation is performed using GC−FID analysis (Mathison and Holstege, 2013).

Chen et al. (2015) detected the presence of brassicasterol in olive oil by GC–Electron Ionization-MS (GC–EI–MS). Grob et al. (1994) were able to detect adulteration of VOO with seed oils (rapeseed, soybean, sunflower and grapeseed) by direct analysis of the sterols using on-line coupled LC–GC–FID.

Steradienes, which are the sterolic degradation products, may be determined either by reversed-phase HPLC making use of their high absorptivity due to conjugation of double bonds, by Gas Liquid Chromatography (GLC) - FID, by GC- MS and, by HPLC coupled on-line to GLC. Determination of sterol isomerization products seems to be the only means to detect an addition of oleic acid rich sunflower seed oil, which has been desterolized, to olive oil (Aladedunye, 2014).

Fourier Transform (FT) -Raman bands due to the major unsaponifiable series of compounds, i.e., squalene, sterolic, and terpenic fractions, analyzed by unsupervised multivariate techniques allow us to differentiate between olive oils and other seed oils as well as and amongst varietal VOO (Baeten, et al., 2001). Near-infrared (IR) proved to be efficient for the detection of sunflower, corn oil and pomace oil in olive oil over a range of

5 to 30% addition (Maggio, et al., 2010). Recently, FT-Raman spectroscopy has been described as a tool for olive oil authentication (Zhang, et al., 2012). Over 90% of the adulterated samples (level of addition 1, 5, and 10%) were correctly classified by applying a set of wavenumbers and using them as input variables for multivariate data evaluation.

Nuclear Magnetic Resonance (NMR) spectroscopy is one of the most promising spectroscopic techniques for the analysis of complex systems, such as food matrices. It has been proven to be very successful in the analysis of olive oil and in combination with multivariate statistical methods was able to classify edible oils and detect olive oil adulteration with seed oils and olive oils of inferior quality (Xu, et al., 2014). It made possible the detection of the presence of refined hazelnut oil in VOO at concentrations 5–10% (Xu, et al., 2014).

Isotope Ratio Mass Spectrometry (IRMS) methods have also been used for the authentication of olive oil by analyzing the bulk oil (Camin, et al., 2016).

CONCLUSION

Adulteration of olive oil is an issue of crucial significance because of its impact in quality, nutritional value and safety of consumers. Due to the inherent EVOO high-cost, the adulteration of this kind of product with low-quality vegetable oils seems to be actually one of the most common types of fraud. The development of analytical methodologies which enable detection of adulterations is warranted since the addition of vegetable oils to EVOO at low percentages could be a very challenging task.

The discovery of reputed markers that could identify possible frauds of EVOO with other vegetable oils could be considered of mandatory importance and will open new avenues in the field of "Food Authenticity". More efforts are needed to exploit new compounds and new methods that could be assigned as reliable adulteration markers able to detect with high selectivity, sensitivity and accuracy blends of EVOO with vegetable oils or olive oils with low grade.

REFERENCES

Aladedunye, F.A. (2014). Natural antioxidants as stabilizers of frying oils. *European journal of lipid science and technology*, 116 : 688-706.

Al-Ismail, K.M., Alsaed, A.K., Ahmad, R. and Al-Dabbas, M. (2010). Detection of olive oil adulteration with some plant oils by GLC analysis of sterols using polar column. *Food Chemistry*, 121: 1255–1259.

Amelio, M., Rizzo, R. and Varazini, F. (1992). Determination of sterols, erythrodiol, uvaol and alkanols in olive oils using combined solid-phase extraction, high-performance liquid chromatography and highresolution gas chromatographic techniques. *Journal of Chromatography* A, 606: 179−185.

Amelotti, G., Daghetta, A. and Ferrario, A. (1989). Content and structure of partial glycerides in virgin olive oil: Their evolution by different working process and preservation form. *Rivista Italiana delle Sostanze Grasse*, 66: 681–692.

Aparicio, R, Morales, M, Aparicio-Ruiz, R, Tena, N, Garcia-Gonzalez, D. (2013). Authenticity of olive oil: mapping and comparing official methods and promising alternatives. *Food Research International*, 54: 2025–2038.

Aparicio, R. and Aparicio-Ruiz, R. (2000). Authentication of vegetable oils by chromatographic techniques. *Journal of Chromatography A*, 881: 93–104.

Aparicio, R. and Luna, G. (2002). Characterisation of monovarietal virgin olive oils. *European Journal of Lipid Science and Technology*, 104: 614–627.

Azadmard-Damirchi, S. and Dutta, P.C. (2007). Free and esterified 4,40-dimethylsterols in hazelnut oil and their retention during refining processes. *Journal of the American Oil Chemists' Society,* 84: 297–304.

Azadmard-Damirchi, S., Savage, G. and Dutta, P. (2005). Sterol fractions in hazelnut and virgin olive oils and 4,4′-dimethylsterols as possible markers for detection of adulteration of virgin olive oil. *Journal of the American Oil Chemists' Society,* 82: 717–725.

Baeten, V., Dardenne, P. and Aparicio, R. (2001). Interpretation of Fourier transform Raman spectra of the unsaponifiable matter in a selection of edible oils. *Journal of Agricultural and Food Chemistry,* 49(11): 5098–5107.

Benitez-Sanchez, P., Leon-Camacho, M. and Aparicio, R. (2003). A comprehensive study of hazelnut oil composition with comparisons to other vegetable oils, particularly olive oil. *European Food Research and Technology,* 218: 13–19.

Bohačenko, I. and Kopicova, Z. (2001). Detection of olive oil authenticity by determination of their sterol content using LC/GC. *Czech Journal of Food Sciences,* 19:97–103.

Busconi, M., Foroni, C., Corradi, M., Bongiorni, C., Cattapan, F. and Fogher, C. (2003). DNA extraction from olive oil and its use in the identification of the production cultivar. *Food Chemistry,* 83:127–34.

Calvano, C.D., DeCeglie, C., D'Accolti, L. and Zambonin, C.G. (2012). MALDI-TOF mass spectrometry detection of extra-virgin olive oil adulteration with hazelnut oil by analysis of phospholipids using an ionic liquid as matrix and extraction solvent. *Food Chemistry,* 134: 1192-1198.

Camin, F., Pavone, A., Bontempo, L., Wehrens, R., Paolini, M., Faberi, A., Marianella, R.M., Capitani, D., Vista, S. and Mannina, L. (2016). The use of IRMS, 1H NMR and chemical analysis to characterise Italian and imported Tunisian olive oils. *Food Chemistry,* 196: 98-105.

Cert, A., Lanzon, A., Carelli, A., Albi, T. and Amelotti, G. (1994). Formation of stigmasta-3,5 diene in vegetable oils. *Food Chemistry,* 49: 287–293.

Chen, Y.Z., Kao, S.Y., Jian, H.C., Yu, Y.M., Li, J.Y., Wang, W.H. and Tsai, C.W. (2015). Determination of cholesterol and four phytosterols in foods without derivatization by gas chromatography-tandem mass spectrometry. *Journal of Food and Drug Analysis*, 23: 636-644.

Commission Regulation (EEC) No. 2568/91 of 11 July 1991. European Communities 1991. *Official Journal of the European Union*, L248.

Elloumi, J., Ben-Ayed, R. and Aifa, S. (2012). An overview of olive oil biomolecules. Current Biotechnology, 1:115–24.

Essid, K., Jahwach-Rabai, W., Trabelsi, M. and Frikha, M. H. (2016). Sterolic Composition of Neutralized Oils Bleached with Clays Activated with Ultrasound. *Iranian Journal of Science and Technology, Transactions A: Science*, 40: 183–189.

Fernández-Cuesta, A., León, L., Velasco, L. and De la Rosa, R. (2013). Changes in squalene and sterols associated with olive maturation. *Food Research International*, 54: 21885-1889.

Fragaki, G., Spyros, A., Siragakis, G., Salivaras, E. and Dais, P. (2005). Detection of extra virgin olive oil adulteration with lampante olive oil and refined olive oil using nuclear magnetic resonance spectroscopy and multivariate statistical analysis. *Journal of Agricultural and Food Chemistry*, 53: 2810–2816.

Frankel, E. (2010). Chemistry of extra virgin olive oil: Adulteration, oxidative stability, and antioxidants. *Journal of Agricultural and Food Chemistry*, 58: 5991– 6006.

Grob, K., Lanfranchi, M. and Mariani, C. (1990) Evaluation of olive oils through the fatty alcohols, the sterols and their esters by coupled LC-GC. *Journal of the American Oil Chemists' Society,* 67: 626–634.

Gunstone, F. (2011). *Vegetable Oils in Food Technology: Composition, Properties and Uses,* 2nd ed.; John Wiley and Sons: West Sussex, U.K.; pp 243−272.

Harwood, J.L. and Aparicio, R. (2000). *Handbook of olive oil: Analysis and properties.* Gaithersburg, MD: Aspen.

International Olive Oil Council (IOOC) (2011). COI/T.15/NC No.3/Rev. 6 November 2011. *Trade standard applying to olive oils and olive-pomace oils,* 1–17.

International Olive Oil Council (IOOC) (2015) COI/T.20/Doc. No. 30/Rev. 1. *Method for the determination of the composition and content of sterols and triterpene dialcohols by capillary column gas chromatography.*

Jabeur, H., Zribi, A., Makni, J., Rebai, A., Abdelhedi, R. and Bouaziz, M. (2014). Detection of Chemlali extra-virgin olive oil adulteration mixed with soybean oil, corn oil, and sunflower oil by using GC and HPLC. *Journal of Agricultural and Food Chemistry*, 62: 4893–4904.

Jimenez de Blas, O. and del Valle Gonzalez, A. (1996). Determination of sterols by capillary column gas chromatography. Differentiation among different types of olive oil: Virgin, refined, and solvent-extracted. *Journal of the American Oil Chemists' Society*, 73: 1685–1689.

Kamm, W., Dionisi, F., Hischenhuber, C. and Engel, K. Authenticity assessment of fats and oils. Food Rev. Int. 2001, 249–290.

Lanzon, A., Cert, A. and Albi, T. (1989). Detection of refined olive oil in virgin olive oil. *Grasas Y Aceites*, 40: 385–388.

Lees, M. (1999). In Food Authenticity: Issues and Methodolo gies, *Eurofins Scientific, Nantes,* 105-1 17.

Lehman, D.W., O'Connor, K., Kovács, B. and Newman, G.E. (2019). Authenticity. *Academy of Management Annals*, Vol. 13, No. 1.

León-Camacho, M., Serrano, M.A. and Constante, E.G. (2004). Formation of stigma-3,5-diene in olive oil during deodorization and/or physical refining using nitrogen as stripping gas. *Grasas y Aceites*, 55: 227–232.

Lukic, M., Lukic, I., Krapac, M., Sladonja, B. and Pilizota, V. (2013) Sterols and triterpene diols in olive oil as indicators of variety and degree of ripening. *Food Chemistry*, 136(1): 251-8.

Maggio, R.M., Cerretani, L., Chiavaro, E., Kaufman, T.S. and Bendini A. (2010). A novel chemometric strategy for the estimation of extra virgin olive oil adulteration with edible oils. *Food Control,* 21: 890-895.

Mariani, C., Bellan, G., Lestini, E. and Aparicio, R. (2006) The detection of the presence of hazelnut oil in olive oil by free and esterified sterols. *European Food Research and Technology*, 223:655–661.

Mathison, B. and Holstege D. (2013). A Rapid Method To Determine Sterol, Erythrodiol, and Uvaol Concentrations in Olive Oil. *Journal of Agricultural and Food Chemistry,* 61: 4506−4513.

Ntsourankoua, T., Artaud, J., Guerere, M. (1994). Triterpene alcohols in virgin olive oil and refined olive pomace oil. *Annales des Falsifications de l'Expertise Chimique et Toxicologique,* 87: 91–107.

Piironen, V., Lindsay, D.G., Miettinen, T.A., Toivo, J. and Lampi, A.M. (2000). Plant sterols: biosynthesis, biological function and their importance to human nutrition. *Journal of the Science of Food and Agriculture,* 80: 939–966.

Rossell, J.B. (2001). Frying: Improving quality. Boca Raton, FL: CRC Press.

Sayago, A, Morales, M.T., Aparicio, R. (2004). Detection of hazelnut oil in virgin olive oil by a spectrofluorimetric method. *European Food Research and Technology,* 218: 480–483.

Srigley, C.T., Oles, C.J., Kia, A.R.F. and Mossoba, M.M. (2016). Authenticity Assessment of Extra Virgin Olive Oil: Evaluation of Desmethylsterols and Triterpene Dialcohols. *Journal of the American Oil Chemists' Society,* 93: 171–181.

Tena, N., Wang, S., Aparicio-Ruiz, R., Garcia-Gonzalez, D. and Aparicio, R. (2015). In-depth assessment of analytical methods for olive oil purity, safety, and quality characterization. *Journal of Agricultural and Food Chemistry,* 63:4509–4526.

Than, T.T., Vergnes, M.F., Kaloustian, J., El-Moselhy, T.F., Amiot-Carlin, M.J. and Portugal, H. Effect of storage and heating on phytosterol concentrations in vegetable oils determined by GC/MS. *Journal of the Science of Food and Agriculture,* 2006: 220−225.

Vekiari, S.A., Oreopoulou, V., Kourkoutas, Y., Kamoun, N., Msallem, M., Psimouli, V., et al. (2010). Characterization and seasonal variation of the quality of virgin olive of the Throumbolia and Koroneiki varieties from Southern Greece. *Grasas y Aceites,* 61: 221–231.

Vichi, S., Pizzale, L., Toffano, E., Bortolomeazzi, R. and Conte, L. (2001). Detection of hazelnut oil in virgin olive oil by assessment of free sterols and triacylglycerols. *Journal of AOAC International,* 84:1534–1541.

Xu, Z., Morris, R.H., Bencsik, M. and Newton M.I. (2014). Detection of Virgin Olive Oil Adulteration Using Low Field Unilateral. NMR. *Sensors*, 14(2): 2028-2035.

Zhang, Q., Liu, C., Sun, Z., Hu, X., Shen, Q. and Wu, J. (2012). Authentication of edible vegetable oils adulterated with used frying oil by Fourier Transform Infrared Spectroscopy. *Food Chemistry*, 132: 1607-1613.

In: Sterols: Types, Classification and Structure ISBN: 978-1-53617-231-7
Editor: Scott Jimenez © 2020 Nova Science Publishers, Inc.

Chapter 7

ANALYSIS OF PHYTOSTEROLS: A SURVEY OF THE ANALYTICAL PROTOCOLS IN USE

Svetlana M. Momchilova[1,*]
and Boryana M. Nikolova-Damyanova[1]
[1]Lab. Chemistry of Lipids, Institute of Organic Chemistry with Centre of Phytochemistry, Bulgarian Academy of Sciences, Sofia, Bulgaria

ABSTRACT

Determination of phytosterols content, including individual components, is, at present, an inevitable part of any intensive research on plants because of their nutritional value or impact on human health. Also, the increased requirements on food quality and authenticity, as well as the expansion of investigations on health beneficial effects of phytosterols, have resulted in the searching and development of efficient analytical methods for their determination. In addition, since each plant has its specific sterol composition, phytosterols can be successfully used as

---

[*] E-mail: svetlana@orgchm.bas.bg.

markers of the authenticity of commercial edible oils and fats, easily revealing any attempt to adulterate food products of animal origin like cheese and butter, or cosmetics.

Analysis of phytosterols is a multistage procedure that includes extraction, isolation/purification as a group of related compounds and a chromatographic technique for separation, identification, and quantification (if required). The main approaches of each step that are widely used at present are presented in this chapter.

Keywords: phytosterols, sterol esters, extraction, analysis, GC, HPLC, MS

INTRODUCTION

Phytosterols, or plant sterols, are natural compounds with, usually, a flat molecule of four rings of 17 carbon atoms, with a hydroxyl group at the 3-position of the A-ring and a side chain of 7 – 10 carbon atoms. The most common phytosterols have a double bond in B-ring with some having a second double bond in the side chain. The saturated sterols are known as stanols. In the nature, phytosterols are present in free form, as esters with fatty acids, as glycosides and acetylated glycosides, the latter being a common form in cereals [1]. The most abundant plant sterols are sitosterol, campesterol, stigmasterol, delta-5- and 7-avenasterol with beta-sitosterol predominating. The sterol composition of nuts, seeds, cereal products, etc. has been reported [2, 3]. Recently, low and very low content of cholesterol was detected in some plant oils as well [4] mostly as fatty acid esters [5]. In addition, some seed oils contain uncommon sterols like, for example, schottenol and spinasterol [6] which can be successfully employed in detection of adulteration [7] and for taxonomic purposes.

The rigid molecular structure of sterols ensures the stability of cell membranes bilayers. In addition, phytosterols have shown to possess various biological activities such as cholesterol-lowering in blood, anti-atherosclerotic, anti-inflammatory, anti-bacterial, antioxidative, anti-cancer properties, etc. Because of their positive health effects, phytosterols are considered as important food components [8]. Mechanism of their biological action and impact have been reviewed [9-11].

Because of their nutritional value and the noted influence on human health, the examination of phytosterols composition and the determination of their content, is, at present, un inevitable part of any intensive research on plants. In addition, the increased requirements on food authenticity and quality, and the increased interest in their beneficial effects on human health, have resulted in searching and development of efficient analytical methods for their determination.

Analysis of phytosterols is a multistage procedure that includes extraction, isolation/purification, separation, identification, and quantification (if required). The main approaches of each step that are widely used at present are presented below.

Additional and more detailed information could be found in some comprehensive reviews [2, 3, 11-13].

EXTRACTION OF PHYTOSTEROLS

Having in mind the complexity of the matrix, determination of natural products, including phytosterols, strongly depends on the choice as well as on the parameters of the extraction procedure and on the way it is carried out [14]. A wrongly chosen isolation may compromise the following and, by far more sophisticated, stages of the analysis.

Sterols, as stated above, are found in free, esterified (with fatty acids) form and as glycosides and acetylated glycosides, the first two classes predominating. The procedure should consider the complexity of the matrix and the phytosterol content and should ensure complete extraction of all sterols in their native form. Note that the phytosterol content in the sample may vary in quite broad limits: from about 300 mg/kg in palm oil to 19000 mg/kg in sesame seed oil, for example [15]. Thus, the choice of all parameters as sample size, solvent(s), temperature, etc., should be made with care and parameters should be optimized in accordance with the aim of the analysis.

The most simple is the solvent-solvent extraction. A comprehensive review [11] presents the widely used conventional Soxhlet extraction and maceration, some nonconventional techniques as microwave-assisted extraction, enzyme-assisted, ultrasonic-assisted, supercritical fluid extractions and some exotic extraction approaches as pressurized liquid extraction and pulsed electric field assisted extraction. It should be stressed that, at present, the simple Soxhlet extraction is the most widely used approach, often serving as a reference method, despite being considered harmful for the health and environment when compared to most of the "nonconventional" techniques.

Soxhlet extraction with hexane, petroleum ether, dichloromethane or ethanol was found suitable to obtain β-sitosterol, campesterol, stigmasterol, fucosterol from leaves, variety of seeds, corn, nuts [11] and from numerous other matrices described in thousands of papers dedicated on the topic (SCOPUS). Another popular method is homogenization of the sample with a single solvent or a solvent mixture, such as chloroform-methanol-water [3] or methylene chloride-isopropanol-methanol [16], for example.

Among the nonconventional approaches, supercritical fluid extraction with CO_2 is the most popular according to [17] and the extraction of β-sitosterol, campesterol, brassicasterol, ergosterol, and stigmasterol from a variety of seeds is described. A recent application of the procedure to determine phytosterols in flaxseed oils is described in [18]. Details on the microwave-assisted extraction of natural compounds, including phytosterols, and its perspectives as a efficient starting procedure can be found in [19].

ISOLATION OF PHYTOSTEROLS AND DETERMINATION OF THEIR CONTENT

Since phytosterols are only minor constituents in plants, extraction usually results in more or less complex mixture of compounds, usually lipids, which often requires additional steps to isolate phytosterols. Chromatography in different modes is the method of choice. Depending on the purpose, at least two preliminary steps are required. To differentiate between the composition of free and esterified phytosterols, the extract is fractionated via any of the chromatographic approaches available in the laboratory. Simple thin layer chromatography on silica gel clearly resolves free sterols from sterol esters which then can be examined separately. For example, single development on a silica gel G 5 x 20 cm plate with mobile phase of hexane-diethyl ether-formic acid (80:20:2 v/v/v) ensures clear separation of sterol esters, triacylglycerols, free fatty acids, free sterols, diacylglycerols, monoacylglycerols and phospholipids (in the order of decreasing Rf value) [20] (Christie and Han 2010, 84). Similar result is achieved by single development with hexane-acetone (100:8 v/v), the only difference being that free sterols migrate ahead of free fatty acids [21].

In most cases the analytical task is to identify and quantify all phytosterols components, ignoring the form in which they are present. In this case, the plant/seed sample should be saponified. Refluxing with 1M potassium hydroxide in 95% ethanol for 1 h, followed by acidifying the water layer with 6M hydrochloric acid is the advised procedure [20] (Christie and Han 2010, 145). However, one should have in mind that saponification is not suitable in case steryl glycosides are present because the acetal bond between the sterol and carbohydrate moieties in the molecule cannot be hydrolyzed in alkaline media [22]. The acid hydrolysis, which is also occasionally employed, may cause decomposition/isomerization of Δ^7-phytosterols during the process [23]. However, since these compounds are rarely present in common seed/plant oils the saponification with ethanolic potassium hydroxide at room or higher temperature is the most frequently used procedure.

Phytosterols remain in the unsaponifiable fraction accompanied by hydrocarbons, long-chain alcohols and other unsaponifiable components of the sample. Therefore, a further stage of purification is required and silica gel column chromatography (CC), thin-layer chromatography (TLC), and solid-phase extraction (SPE) are usually employed for the purpose using suitable elution solvent mixtures most often of hexane or petroleum ether with diethyl ether or acetone [3, 11, 24]. SPE is considered better for routine analyses because it is time-saving and provides reproducible elution. SPE is performed successfully in different modes: normal phase on silica or alumina, or reversed phase on C18 silica, and there is a large number of papers describing the successful application of any of these. Normal phase SPE protocols are described for isolation of phytosterols converted into trimethylsilyl ether derivatives [25], and for the separate isolation of free sterols and steryl glycosides [23, 26]. Good example of SPE on neutral alumina is the separation of free and esterified sterols described in [27]. A C18 reversed phase SPE was used for isolation of sterols from unsaponifiables [28].

DETERMINATION OF PHYTOSTEROLS COMPOSITION

Chromatography is not the single method employed in the analysis of phytosterols at present, but combined with the modern powerful detection techniques, it is undoubtedly the most widely used approach for their identification and quantification.

Gas Chromatography

Gas chromatography (GC) is a good example of the above statement since it is still the most widely used method for separation, identification, and quantification of phytosterols [12, 29, 30]. Capillary columns with non-polar phase (100% cross-linked polysiloxane) are used by far more often than the polar phases (phenyl- or diphenyl-dimethyl-polysiloxanes, cyanopropyl-phenyl-methylpolysiloxanes, etc.) although the latter show higher thermal stability [2, 3, 11, 31-33]. The protocol usually includes derivatization of phytosterols to the highly volatile trimethylsilyl ethers considered better for GC analysis.

Several reagents can be used for the purpose, i.e., N-methyl-N-(trimethylsilyl)-trifluoroacetamide in anhydrous pyridine, bis(trimethylsilyl)-trifluoroacetamide with 1% trimethylchlorosilane (1:1v/v), and occasionally hexamethyldisilazane with dry pyridine or trimethylchlorosilane. The sililated derivatives are reported to ensure better peak shapes and better resolution of individual phytosterols and phytostanols [33]. However, our own experience and the results reported by others show that GC of underivatized sterols provides equally good results and in addition, by omitting the additional derivatization step, the sources of experimental errors and the loss of components are minimized [5, 21, 31, 34-36].

The GC quantitation of sterols includes usually the employment of an internal standard. Various compounds are suitable for the purpose but most popular are 5α-cholestane, betulin, 5β-cholestan-3α-ol (epicoprostanol) or 5α-cholestan-3β-ol [3]. The routine GC quantitative analysis of sterols uses flame ionization detector (FID) and identification of peaks is performed by comparing the respective retention times with these of individual reference sterols. However, unambiguous identification of peaks is performed by using mass spectrometric (MS) detector [37]. It has been reported that using of sterol derivatives such as acetates, trifluoroacetates, methyl ethers, trimethylsilyl ethers, picolinyl esters, N-methylpyridyl ethers or sulphated esters increases by far the sensitivity of detection. Comprehensive data on GC-MS of sterols and the characteristic mass fragmentations of the most abundant components are given in [38]. Additional data on employing GC and GC-MS in the analysis of sterols in a variety of plants via various chromatographic conditions can be found in [2, 3, 11, 13].

High Performance Liquid Chromatography

Since the late 1970ties, High-Performance Liquid Chromatography (HPLC) is intensively employed in the analysis of phytosterols. HPLC is considered advantageous because it is carried out at mild experimental conditions and, depending on the detection mode chosen, it could be a non-destructive method providing pure phytosterol fraction or individual component for further structural analysis. There is a large, although nonsignificant variation in HPLC analytical protocols in recent years due to the very large number of plants examined (including very exotic) and to the increased interest in phytosterol composition and in their beneficial effect on the human health. In general, mobile phases and elution mode are chosen with care, firstly, for ensuring the desired good resolution for a reasonable time and secondly, to meet the limitation and requirements of the detection mode. In the case of relatively simple matrix composition, preliminary purification of the plant extract might be not needed. Both normal phase HPLC (for resolving the main phytosterols classes, see [39]) and reversed-phase HPLC (on C8 or C18 silica columns, for separation, identification and quantification of individual components) have been employed. Normal phase HPLC uses gradient elution with hexane-based mobile phases and acetone or isopropanol as a polar modifier. Mobile phases based mostly on methanol and/or acetonitrile and butanol modified with water are in use with reversed-phase HPLC.

Detectors differ as well: ultraviolet (UV, at 200 nm usually), refractive index (RI), photodiode array (PDA), evaporative light scattering detector (ELSD). Note, that using UV detection at 200 nm limits the choice of mobile phase solvents, excluding these with a cut-off above this wavelength, like acetone, for example. A fine recent example of using UV detection is described in [40] and that of using PDA: in [41].

HPLC coupled with mass detector (HPLC-MS) gains significant attention at present, mostly because it allows for unequivocal identification of each component (as in GC-MS) and because of the low and very low detection limit.

In addition, due to the specific fragmentation: intense fragments at m/z 369, 395, 397, 383, 379.3 in selected ion monitoring mode (SIM) for cholesterol, stigmasterol, β-sitosterol, campesterol, and ergosterol, respectively (assigned to the loss of water), identification is possible even if the species are not fully resolved [3, 11, 13].

Of the various MS techniques, atmospheric pressure chemical ionization (APCI) is most widely used, while any of the electron impact (EI), electrospray ionization (ESI) and atmospheric pressure photoionization (APPI) could be an effective alternative.

Consult the comprehensive recent review [13] for details in the application of mass spectrometry coupled with GC and HPLC in the analysis of phytosterols.

Nuclear Magnetic Resonance Spectrometry

Nuclear magnetic resonance spectrometry (NMR) in ^{1}H-NMR mode is used relatively rarely at present mostly as an approach to identify complex phytosterol components in different samples. NMR could be applied as a stand-alone method, directly on the total sample extract as shown in [42-44] or as coupled with HPLC separation [11]. The potential of ^{13}C NMR, one-dimensional (1D) and two-dimensional (2D) NMR techniques for structure elucidation and signal assignments of sterols and related compounds has been discussed in detail in [45-47].

PHYTOSTEROLS AS MARKERS FOR AUTHENTICITY OF EDIBLE FATS AND OILS

In addition to all the benefits as nutraceuticals, phytosterols find recently another important application. The years-long examination of phytosterols composition has shown that it differs in different plants, in different varieties of the same plant, and depends strongly on the geographic location and climate changes. Phytosterols are therefore a promising marker for determining the authenticity of a given plant product (oil mostly) or might help to trace adulteration in expensive oils like olive [48, 49], pumpkin [50] or argan oil [7], for example. In most cases a thorough preliminary work is required in investigating a large number of samples of the same plant obtaining reliable qualitative, and, especially, quantitative data on the phytosterol composition, and thus, allowing to distinguish clearly between different cultivars and geographical origin [51, 52]. These data, combined with statistical methods, allow for trustworthy authentication of the oils and for tracing of adulteration. For example, highly sophisticated techniques like ultra-high-pressure liquid chromatography (UHPLC) coupled to quadrupole-time-of-flight mass spectrometry (QTOF-MS) plus multivariate statistics were applied to discriminate between extra virgin olive oils (EVOO, the most expensive olive oil, considered as most beneficial to the human health, as well) originated from 6 Italian regions [52]. Similar approach was applied to discriminate between six varieties of grapes [53] which might help to prove the authenticity of expensive vines. A combination of GC-FID, NMR, and chemometric approaches were applied to determine the phytosterols content (total content and the content of each component: cholesterol, brassicasterol, campesterol, stigmasterol, β-sitosterol and Δ^7-stigmasterol) in olive oils from a specific Spanish region [54].

Phytosterols appeared to be useful markers in tracing adulteration of animal edible fats, like butter and cheese [55], lard and tallow [56] with plant oils, as well.

Also, phytosterols were found to be reliable biomarkers, better than fatty acids composition, in chemotaxonomy of phytoplankton after GC-MS determination of their silylated derivatives [57].

The analysis of phytosterols was, and still is, a challenge for the analysts. At present, this is in most cases a multistage procedure and none of the sophisticated techniques for separation and identification of the components can replace the precisely performed isolation of these minor compounds from the complex natural matrices. Further methodological development is however inevitable because of the increasing interest in phytosterols due to their beneficial effect on human health.

REFERENCES

[1] Moreau, R. A., Whitaker, B. D. and Hicks, K. B. 2002. "Phytosterols, phytostanols, and their conjugates in foods: structural diversity, quantitative analysis, and health-promoting uses". *Progress in Lipid Research*, 41:457 - 500.

[2] Abidi, S. L. 2001. "Chromatographic analysis of plant sterols in foods and vegetable oils". *Journal of Chromatography A*, 935:173 - 201.

[3] Lagarda, M. J., García-Llatas, G. and Farré, R. 2006. "Analysis of phytosterols in foods". *Journal of Pharmaceutical and Biomedical Analysis,* 41:1486 - 96.

[4] Schwartz, Heidi, Ollilainen, Velimatti, Piironen, Vieno and Lampi, Anna-Maija. 2008. "Tocopherol, tocotrienol and plant sterol contents of vegetable oils and industrial fats". *Journal of Food Composition and Analysis,* 21:152 - 61.

[5] Taneva, Sabina, Momchilova, Svetlana, Marekov, Ilko, Blagoeva, Elitsa and Nikolova, Magdalena. 2013. "Free and Esterified Sterols in Walnuts and Hazelnuts in Three Stages During Kernel Development". *Comptes rendus de l'Academie bulgare des Sciences,* 66:1681 - 8.

[6] El Kharrassi, Youssef, Samadi, Mohammad, Lopez, Tatiana, Nury, Thomas, El Kebbaj, Riad, Andreoletti, Pierre, El Hajj, Hammam I., Vamecq, Joseph, Moustaid, Khadija, Latruffe, Norbert, El Kebbaj, M'Hammed Saïd, Masson, David, Lizard, Gérard, Nasser, Boubker and Cherkaoui-Malki, Mustapha. 2014. "Biological activities of Schottenol and Spinasterol, two natural phytosterols present in argan oil and in cactus pear seed oil, on murine miroglial BV2 cells". *Biochemical and Biophysical Research Communications,* 446:798 - 804.

[7] Momchilova, S. M., Taneva, S. P., Dimitrova, R. D., Totseva, I. R. and Antonova, D. V. 2016. "Evaluation of authenticity and quality of Argan oils sold on the Bulgarian market". *La Rivista Italiana delle Sostanze Grasse,* XCIII:95 - 103.

[8] Dutta, Paresh C. 2004. *Phytosterols as Functional Food Components and Nutraceuticals.* New York, Basel: Marcel Dekker.

[9] Tikkanen, M. J. 2005. "Plant Sterols and Stanols". *Handbook of Experimental Pharmacology,* 170:215 - 30.

[10] Gupta, A. K., Savopoulos, C. G., Ahuja, J. and Hatzitolios, A. I. 2011. "Role of phytosterols in lipid-lowering: current perspectives". *QJMed,* 104:301 - 8.

[11] Uddin, M. S., Sahena, Ferdosh, Jahurul, Akanda Haque, Kashif, Ghafoor, Rukshana, A. H., Eaqub, Ali, Kamaruzzaman, B. Y., Fauzi,

M. B., Hadijah, S., Sharifudin, Shaarani and Zaidul, Sarker Islam. 2018. "Techniques for the extraction of phytosterols and their benefits in human health: a review". *Separation Science and Technology*, 53:2206 - 23.

[12] Bernal, J., Mendiola, J. A., Ibáñez, E. and Cifuentes, A. 2011. "Advanced analysis of nutraceuticals". *Journal of Pharmaceutical and Biomedical Analysis*, 55:758 - 74.

[13] Gachumi, George and El-Aneed, Anas. 2017. "Mass Spectrometric Approaches for the Analysis of Phytosterols in Biological Samples". *Journal of Agricultural and Food Chemistry*, 65:10141 - 56.

[14] Smith, Roger M. 2003. "Before the injection – modern methods of sample preparation for separation techniques". *Journal of Chromatography A*, 1000:3 - 27.

[15] Codex Standard for Named Vegetable Oils (CODEX-STAN 210-1999).

[16] Chakrabarti, Priyadarshini, Morré, Jeffery T., Lucas, Hannah M., Maier, Claudia S. and Sagili, Ramesh R. 2019. "The omics approach to bee nutritional landscape". *Metabolomics*, 15:127(1 - 10).

[17] Herrero, Miguel, Mendiola, Jose A., Cifuentes, Alejandro and Ibaneza, Elena. 2010. "Supercritical fluid extraction: Recent advances and applications". *Journal of Chromatography A*, 1217:2495 - 2511.

[18] Dąbrowski, Grzegorz, Czaplicki, Sylwester and Konopka, Iwona. 2019. "Fractionation of sterols, tocols and squalene in flaxseed oils under the impact of variable conditions of supercritical CO_2 extraction". *Journal of Food Composition and Analysis*, 83:103261.

[19] Chan, Chung-Hung, Yusoff, Rozita, Ngoh, Gek-Cheng and Kung, Fabian Wai-Lee. 2011. "Microwave-assisted extractions of active ingredients from plants". *Journal of Chromatography A*, 1218: 6213 - 25.

[20] Christie, William W. and Xianlin Han. 2010. *Lipid Analysis: Isolation, Separation, Identification and Lipidomic Analysis* (Fourth Edition). Bridgwater, England: The Oily Press.

[21] Zlatanov, Magdalen, Antova, Ginka, Angelova-Romova, Maria, Momchilova, Svetlana, Taneva, Sabina and Nikolova-Damyanova, Boryana. 2012. "Lipid Structure of Lallemantia Seed Oil - a Potential Source of omega-3 and omega-6 Fatty Acids for Nutritional Supplements". *Journal of the American Oil Chemists Society,* 89:1393 - 1401.

[22] Heupel, R. C. 1989. "Isolation and primary characterization of sterols". In *Analysis of sterols and Other Biologically Significant Steroids,* edited by Ness, W. D. and E. J. Parish, 49 - 60. San Diego, California: Academic Press Inc.

[23] Breinhoelder, P., Mosca, L. and Lindner, W. 2002. "Concept of sequential analysis of free and conjugated phytosterols in different plant matrices". *Journal of Chromatography B,* 777:67 - 82.

[24] Azadmard-Damirchi, S. 2010. "Solid-phase extraction as an alternative to thin-layer chromatography in phytosterol analysis". In *Chromatography: Types, Techniques and methods,* edited by T. J. Quintin, 389 - 404. New York: Nova Science Publishers.

[25] Toivo, J., Piironen, V., Kalo, P. and Varo, P. 1998. "Gas chromatographic determination of major sterols in edible oils and fats using solid-phase extraction in sample preparation". *Chromatographia,* 48:745 - 50.

[26] Parcerisa, J., Richardson, D. G., Rafecas, M., Codony, R. and Boatella, J. 1998. "Fatty acid, tocopherol and sterol content of some hazelnut varieties (Corylus avellana L.) harvested in Oregon (USA)". *Journal of Chromatography A,* 805:259 - 68.

[27] Phillips, K. M., Ruggio, D. M., Toivo, J. I., Swank, M. A. and Simpkins, A. H. 2002. "Free and esterified sterol composition of edible oils and fats". *Journal of Food Composition and Analysis,* 15:123 - 42.

[28] Ham, B., Butler, B. and Thionville, P. 2000. "Evaluating the isolation and quantification of sterols in seed oils by solid-phase extraction and capillary gas-liquid chromatography". *LC-GC North America,* 18:1174 - 81.

[29] Du, M. and Ahn, D. U. 2002. "Simultaneous Analysis of Tocopherols, Cholesterol, and Phytosterols Using Gas Chromatography". *Journal of Food Science,* 67:1696 - 1700.

[30] Fanali, Chiara, Dugo, Laura, Dugo, Paola and Mondello, Luigi. 2013. "Capillary-liquid chromatography (CLC) and nano-LC in food analysis". *Trends in Analytical Chemistry,* 52:226 - 38.

[31] Momchilova, Svetlana, Antonova, Daniela, Marekov, Ilko, Kuleva, Liliana and Nikolova-Damyanova, Boryana. 2007. "Fatty Acids, Triacylglycerols, and Sterols in Neem Oil (Azadirachta Indica A. Juss) as Determined by a Combination of Chromatographic and Spectral Techniques". *Journal of Liquid Chromatography and Related Technologies,* 30:11 - 25.

[32] Dutta, P. C. and Normén, L. 1998. "Capillary column gas-liquid chromatographic separation of Δ^5-unsaturated and saturated phytosterols". *Journal of Chromatography A,* 816:177 - 84.

[33] Laakso, P. 2005. "Analysis of sterols from various food matrices". *European Journal of Lipid Science and Technology,* 107:402 - 10.

[34] Zlatanov, M. D., Angelova-Romova, M. J., Antova, G. A., Dimitrova, R. D., Momchilova, S. M. and Nikolova-Damyanova, B. M. 2009. "Variations in Fatty Acids, Phospholipids and Sterols During the Seed Development of a High Oleic Sunflower Variety". *Journal of the American Oil Chemists Society,* 86:867 - 75.

[35] Chen, Yan-Zong, Kao, Shih-Yao, Jian, Hao-Cheng, Yu, Yu-Man, Lia, Ju-Ying, Wang, Wei-Hsien and Tsai, Chung-Wei. 2015. "Determination of cholesterol and four phytosterols in foods without derivatization by gas chromatography-tandem mass spectrometry". *Journal of Food and Drug Analysis,* 23:636 - 44.

[36] Huang, Chin-Yuan, Yung-Lin, Chu, Kandi, Sridhar and Pi-Jen, Tsai. 2019. "Analysis and determination of phytosterols and triterpenes in different inbred lines of Djulis (Chenopodium formosanum Koidz) hull: A potential source of novel bioactive ingredients". *Food Chemistry,* 297:124948 (1 - 7).

[37] Matysik, S., Klunemann, H. H. and Schmitz, G. 2012. "Gas Chromatography–Tandem Mass Spectrometry Method for the

Simultaneous Determination of Oxysterols, Plant Sterols, and Cholesterol Precursors". *Clinical Chemistry,* 58:1557 - 64.

[38] Goad, L. John and Toshihiro Akihisa. 1997. "Mass spectrometry of sterols". In *Analysis of Sterols,* 152 - 96. Dordrecht: Springer.

[39] Nestola, Marco and Schmidt, Torsten C. 2016. "Fully automated determination of the sterol composition and total content in edible oils and fats by online liquid chromatography–gas chromatography–flame ionization detection". *Journal of Chromatography A,* 1463: 136 - 43.

[40] Talreja, Tamanna, Kumar, Mangesh, Goswami, Asha, Gahlot, Ghanshyam, Jinger, Surendra Kumar and Sharma, Tribhuwan. 2017. "Qualitative and quantitative estimation of phytosterols in Achyranthes aspera and Cissus quadrangularis by HPLC". *The Pharma Innovation Journal,* 6:76 - 9.

[41] Nachimuthu, Kannikaparameswari and Palaniswamy, Chinnaswamy. 2013. "Physiochemical analysis and HPLC-PDA method, for quantification of stigmasterol in Dipteracanthus patulus (Jacq.) Nees". *Journal of Pharmacy Research,* 7:741 - 6.

[42] Sopelana, P., Ibargoitia, María L. and Guillén, María D. 2016. "Influence of fat and phytosterols concentration in margarines on their degradation at high temperature. A study by ^{1}H Nuclear Magnetic Resonance". *Food Chemistry,* 197:1256 - 63.

[43] Zhang, X.-L., Wang, C., Chen, Z., Zhang, P.-Y. and Liu, H.-B. 2016. "Development and Validation of Quantitative ^{1}H NMR Spectroscopy for the Determination of Total Phytosterols in the Marine Seaweed Sargassum". *Journal of Agricultural and Food Chemistry,* 64: 6228 - 32.

[44] Chundattu, Sony J., Agrawal, Vijay Kumar and Ganesh, N. 2016. "Phytochemical investigation of *Calotropis procera*". *Arabian Journal of Chemistry,* 9:S230 - 4.

[45] Goad, L. John and Toshihiro Akihisa. 1997. "^{1}H NMR spectroscopy of sterols". In *Analysis of Sterols,* 197 - 234. Dordrecht: Springer.

[46] Goad, L. John and Toshihiro Akihisa. 1997. "^{13}C NMR spectroscopy of sterols". In *Analysis of Sterols,* 235 - 55. Dordrecht: Springer.

[47] Goad, L. John and Toshihiro Akihisa. 1997. "One-dimensional and two-dimensional NMR spectroscopy of sterols". In *Analysis of Sterols*, 256 - 76. Dordrecht: Springer.

[48] Azadmard-Damirchi, S. and Torbati, M. 2015. "Adulterations in Some Edible Oils and Fats and Their Detection Methods". *Journal of Food Quality and Hazards Control*, 2:38 - 44.

[49] Al-Ismail, Khalid M., Alsaed, Ali K., Ahmad, Rafat and Al-Dabbas, Maher. 2010. "Detection of olive oil adulteration with some plant oils by GLC analysis of sterols using polar column". *Food Chemistry*, 121:1255 - 9.

[50] Dulf, Francisc Vasile, Bele, Constantin, Unguresan, Mihaela, Parlog, Raluca and Socaciu, Carmen. 2009. "Phytosterols as Markers in Identification of the Adulterated Pumpkin Seed Oil with Sunflower Oil". *Bulletin UASVM Agriculture*, 66:301 - 7.

[51] Olmo-García, Lucía, Polari, Juan J., Li, Xueqi, Bajoub, Aadil, Fernández-Gutiérrez, Alberto, Wang, Selina C. and Carrasco-Pancorbo, Alegría. 2019. "Study of the minor fraction of virgin olive oil by a multi-class GC–MS approach: Comprehensive quantitative characterization and varietal discrimination potential". *Food Research International*, 125:108649.

[52] Ghisoni, Silvia, Lucini, Luigi, Angillett, Federica, Rocchetti, Gabriele, Farinelli, Daniela, Tombesi, Sergio and Trevisan, Marco. 2019. "Discrimination of extra-virgin-olive oils from different cultivars and geographical origins by untargeted metabolomics". *Food Research International*, 121:746 - 53.

[53] Millán, Laura, Sampedro, M. Carmen, Sánchez, Alicia, Delporte, Cédric, Van Antwerpen, Pierre, Goicolea, M. Aranzazu and Barrio, Ramón J. 2016. "Liquid chromatography–quadrupole time of flight tandem mass spectrometry–based targeted metabolomic study for varietal discrimination of grapes according to plant sterols content". *Journal of Chromatography A*, 1454:67 - 77.

[54] Sayagoa, Ana, González-Domíngueza, Raúl, Urbano, Juan and Fernández-Recamales, Ángeles. 2019. "Combination of vintage and new-fashioned analytical approaches for varietal and geographical

traceability of olive oils". *LWT - Food Science and Technology,* 111:99 - 104.

[55] Kim, Nam Sook, Lee, Ji Hyun, Han, Kyoung Moon, Kim, Ji Won, Cho, Sooyeul and Kim, Jinho. 2014. "Discrimination of commercial cheeses from fatty acid profiles and phytosterol contents obtained by GC and PCA". *Food Chemistry,* 143:40 - 7.

[56] Liao, Chia-Ding, Peng, Guan-Jhih, Ting, Yueh, Chang, Mei-Hua, Tseng, Su-Hsiang, Kao, Ya-Min, Lin, King-Fu, Chiang, Yu-Mei, Yeh, Ming-Kung and Cheng, Hwei-Fang. 2017. "Using phytosterol as a target compound to identify edible animal fats adulterated with cooked oil". *Food Control,* 79:10 - 6.

[57] Taipale, Sami J., Hiltunen, Minna, Vuorio, Kristiina and Peltomaa, Elina. 2016. "Suitability of Phytosterols Alongside Fatty Acids as Chemotaxonomic Biomarkers for Phytoplankton". *Frontiers in Plant Science,* 7:212.

BIOGRAPHICAL SKETCH

Svetlana Momchilova

Affiliation: Institute of Organic Chemistry with Centre of Phytochemistry, Bulgarian Academy of Sciences

Education: M.S. in Chemistry, Sofia University St. Kliment Ohridski, Faculty of Chemistry (1989);
PhD in Chemistry (Bioorganic Chemistry, Chemistry of Natural and Physiologically Active Substances), Bulgarian Academy of Sciences, Institute of Organic Chemistry with Centre of Phytochemistry (2002).

Business Address:
Lab. Chemistry of Lipids, Institute of Organic Chemistry with Centre of Phytochemistry, Bulgarian Academy of Sciences, Acad. George Bonchev Str., block 9, 1113 Sofia, Bulgaria

Research and Professional Experience: research work in the field of lipids, chromatographic analysis of lipids, fatty acids, acylglycerols, sterols, neutral lipid classes, isomeric fatty acids, isomeric triacylglycerols, quality and authenticity of food, food supplements, plant biologically active substances, antioxidants, spectroscopy, microelements; coordinator of scientific projects; supervisor of MS students; author and co-author of 70 scientific articles with >535 citations.

Professional Appointments:
Since 2007: Associate Professor in Bioorganic Chemistry, Chemistry of Natural and Physiologically Active Substances, Institute of Organic Chemistry with Centre of Phytochemistry, Bulgarian Academy of Sciences; since 2010: Head of the Laboratory Chemistry of Lipids, Institute of Organic Chemistry with Centre of Phytochemistry, Bulgarian Academy of Sciences

Honors: "Best scientific paper" chosen by the Colloquium „Chemistry of Natural Products" – Institute of Organic Chemistry with Centre of Phytochemistry-Bulgarian Academy of Sciences: 2005, 2006, 2010

Publications from the Last 3 Years: 2016-2019

[1] Momchilova, Sv., Arpadjan, S. and Blagoeva, E. 2016. "Accumulation of Microelements (Cd, Cu, Fe, Mn, Pb, Zn) in Walnuts (*Juglans regia* L.) from Bulgaria Depending on the Cultivar and the Harvesting Year". *Bulgarian Chemical Communications,* 48 (1):50 - 4.

[2] Momchilova, S. M., Taneva, S. P., Dimitrova, R. D., Totseva, I. R. and Antonova, D. V. 2016. "Evaluation of authenticity and quality of argan oils sold on the Bulgarian market". *Rivista Italiana delle Sostanze Grasse,* XCIII (Apr.-Jun.):95 - 103.

[3] Momchilova, Sv. and Nikolova-Damyanova, B. 2016. "Silver-ion Chromatography of Fatty Acids". In *Encyclopedia of Lipidomics*,

Editor Marcus R. Wenk, Springer, DOI: 10.1007/978-94-007-7864-1_75-1 (ISBN 978-94-007-7864-1) http://link.springer.com/referenceworkentry/10.1007/978-94-007-7864-1_75-1.

[4] Taneva, S., Konakchiev, A., Totzeva, I., Kamenova-Nacheva, M., Nikolova, Y., Momchilova, S. and Dimitrov, V. 2017. "Super-critical carbon dioxide extraction as an effective green technology for production of high quality rose hip oil". *Bulgarian Chemical Communications,* 49(Special Edition B):126 - 31.

[5] Momchilova, S. M., Taneva, S. P., Zlatanov, M. D., Antova, G. A., Angelova-Romova, M. J. and Blagoeva, E. 2017. "Fatty acids, tocopherols and oxidative stability of hazelnuts during storage". *Bulgarian Chemical Communications,* 49(Special issue G):65 - 70.

[6] Yordanova, V., Momchilova, Sv., Momchilova, A., Ivanova, A. and Maslenkova, L. 2017. "Influence of abiotic environmental factors on photosynthetic activity and leaf fatty acid composition of common butterbur (*Petasites hybridus*) from different habitats". *Comptes rendus de l'Academie bulgare des Sciences,* 70(10):1399 - 404.

[7] Angelova, L., Ivanova, A., Momchilova, Sv., Momchilova, A. and Maslenkova, L. 2017. "Functional characteristics of photosynthetic apparatus of *Lactuca tatarica* (Asteraceae) – a halophyte plant from Black sea coast". *Comptes rendus de l'Academie bulgare des Sciences*, 70(10):1405 - 10.

[8] Mutafova, B., Momchilova, S., Pomakova, D., Avramova, T. and Mutafov, S. 2018. "Enhanced cell surface hydrophobicity favors the 9α-hydroxylation of androstenedione by resting Rhodococcus sp. Cells". *Engineering in Life Sciences,* 18(12):949 - 54.

[9] Simonovska, J. M., Yancheva, D. Y., Mikhova, B. P., Momchilova, S. M., Knez, Ž. F., Primožić, M. J., Kavrakovski, Z. S. and Rafajlovska, V. G. 2019. "Characterization of extracts from red hot pepper (*Capsicum annuum* L.)". *Bulgarian Chemical Communications,* 51(1):103 - 12.

[10] Momchilova, Sv. M., Taneva, S. P., Totseva, I. R., Nikolova, Y. I., Karakirova, Y. G., Aleksieva, K. I., Mladenova, R. B. and Kancheva, V. D. 2019. "Gamma-irradiation of nuts – EPR characterization and

effects on lipids and oxidative stability: I. Hazelnuts". *Bulgarian Chemical Communications,* 51(Special Issue A):256 - 262.

[11] Momchilova, Sv. M., Taneva, S. P., Totseva, I. R., Nikolova, Y. I., Karakirova, Y. G., Aleksieva, K. I., Mladenova, R. B. and Kancheva, V. D. 2019. "Gamma-irradiation of nuts – EPR characterization and effects on lipids and oxidative stability: II. Peanuts". *Bulgarian Chemical Communications,* 51(Special Issue A):263 - 9.

[12] Denev, P., Klisurova, D., Teneva, D., Ognyanov, M., Georgiev, Y., Momchilova, S. and Kancheva, V. 2019. "Effect of gamma irradiation on the chemical composition and antioxidant activity of dried black chokeberry (*Aronia melanocarpa*) fruits". *Bulgarian Chemical Communications,* 51(Special Issue A):270 - 5.

[13] Karamalakova, Yanka D., Nikolova, Galina D., Denev, Petko N., Momchilova, Svetlana, Slavova-Kazakova, Adriana K., Kancheva, Vesela D., Zheleva Antoaneta M. and Gadjeva, Veselina G. 2019. "High-level gamma radiation effects on radical-scavenging activity of Black Chokeberry (Aronia melanocarpa) ethanol extract". *Bulgarian Chemical Communications,* 51(Special Issue A):276 - 82.

[14] https://www.scopus.com/authid/detail.uri?authorId=6602893935.

[15] https://www.scopus.com/authid/detail.uri?authorId=55387026100.

[16] https://www.scopus.com/authid/detail.uri?authorId=23068235200&eid=2-s2.0-0344443922.

[17] https://www.scopus.com/authid/detail.uri?authorId=6504266742&eid=2-s2.0-17044458907.

INDEX

#

24-nordehydrocholesterol, 19, 38
4,24 dimethyl, 27, 28, 29
5,7 dien-3β-ol, 28, 29

A

accessions, 84
acetone, 25, 121, 124
acid, x, 3, 4, 6, 13, 38, 40, 45, 62, 71, 73, 76, 96, 101, 118, 121, 129, 133, 135
adsorption, 47, 53
adulteration, xi, 61, 62, 69, 77, 86, 87, 97, 98, 99, 100, 103, 104, 105, 106, 107, 108, 109, 110, 111, 112, 113, 114, 116, 118, 125, 126, 132
alcohols, 2, 3, 6, 63, 68, 75, 101, 102, 109, 113, 115, 121
algae, 4, 21, 22, 25, 28, 29, 33, 37, 38
alkaloids, 45, 46
Amazonia, 44, 55
AMDIS, 17
amines, 3
amino acids, 6
analysis, viii, x, xi, 18, 19, 39, 40, 41, 62, 63, 64, 65, 68, 69, 70, 74, 76, 77, 78, 81, 83, 84, 88, 89, 90, 91, 93, 95, 98, 101, 105, 109, 110, 111, 112, 113, 118, 119, 122, 123, 124, 126, 128, 129, 130, 131, 132, 134
aniline, 99
antioxidant, 86, 87, 95, 106, 136
Aplysina sp, 28, 29, 30, 31, 33, 34, 37
argon, 6
assessment, 84, 103, 114, 115
atmospheric pressure, 124
authentication, 60, 87, 100, 110, 125
authenticity, vi, xi, xii, 61, 66, 68, 78, 79, 80, 89, 91, 97, 98, 99, 100, 102, 103, 105, 108, 110, 111, 112, 114, 115, 117, 119, 125, 127, 134
avenasterol, ix, x, 60, 61, 63, 64, 66, 67, 68, 69, 74, 78, 79, 80, 81, 82, 89, 90, 91, 92, 102, 104, 105, 106, 107, 118

B

B. rufus, 29, 32, 33, 34
bacteria, 22
Baikal lake, 44

base, 104, 109, 132
beneficial effect, x, xii, 85, 86, 87, 117, 119, 124, 126
benefits, vii, ix, 59, 78, 98, 99, 125, 128
benzene, 3
biochemistry, 56
biological activities, 46, 55, 118
biomarker, 28, 29, 33, 35, 37
biomarkers, 4, 22, 24, 25, 28, 29, 34, 37, 41, 45, 126, 133
biomolecules, 3, 113
biosynthesis, 37, 40, 46, 56, 115
biotechnology, 41
bleaching, 108
blends, xi, 98, 105, 111
brassicastanol, 20, 28
brassicasterol, 26, 30, 32, 35, 104
breast cancer, 95, 96
bronchial epithelial cells, 96
brown alga, 4, 25, 28, 29, 33, 38
brown algae, 4, 25, 28, 29, 33, 38

C

C. personatus, 29, 32, 34
calibration, 18, 19
calibration curve, 18, 19
campesterol, ix, x, xi, 5, 18, 20, 25, 27, 30, 32, 60, 61, 63, 64, 66, 67, 68, 69, 71, 74, 78, 79, 80, 81, 82, 84, 86, 89, 90, 91, 98, 102, 103, 104, 106, 107, 108, 118, 120, 124, 126
capillary, 8, 65, 77, 78, 88, 109, 114, 129
Caranx hippos, viii, 2, 24, 32, 34
carbon, viii, 2, 8, 13, 15, 38, 39, 55, 118, 135
cardiovascular disease, 86
categorization, 99
cell membranes, 2, 5, 118
cell surface, 135
challenges, 108

characteristic ions, 6, 16
cheese, xii, 118, 126
chemical, ix, 6, 43, 45, 46, 60, 63, 68, 76, 95, 105, 112, 124, 136
chemical characteristics, 95
cholestanol, 20, 28, 65, 77, 89
cholesterol, ix, x, 3, 4, 5, 13, 15, 18, 20, 21, 22, 25, 26, 28, 30, 32, 33, 34, 36, 38, 40, 46, 50, 53, 60, 61, 63, 66, 67, 69, 78, 79, 80, 81, 83, 85, 89, 90, 91, 100, 102, 103, 112, 118, 124, 126, 130, 131
cholesterol lowering agents, 83
chromatographic conditions, 7, 18, 123
chromatographic technique, viii, xii, 51, 52, 69, 83, 111, 118
chromatography, vii, viii, 1, 6, 40, 41, 47, 48, 51, 52, 53, 65, 77, 78, 88, 113, 114, 121, 122, 129, 130, 131, 132
clams, 24, 30
classes, 7, 35, 37, 39, 44, 101, 119, 124, 134
classification, 62, 68, 100
coenzyme, 3
collateral, 5
colon, x, 85, 90
colon cancer, x, 85
colonization, 44
commercial, xii, 17, 18, 84, 99, 100, 106, 118, 133
commodity, 99, 104
community, viii, 43, 105
compliance, 66
composition, ix, x, xii, 22, 26, 30, 32, 43, 45, 46, 55, 60, 61, 62, 63, 66, 68, 69, 70, 71, 73, 76, 78, 80, 81, 82, 83, 84, 86, 87, 89, 90, 91, 93, 94, 95, 100, 102, 103, 104, 105, 107, 109, 112, 113, 117, 118, 119, 121, 123, 125, 126, 129, 131, 135, 136
compounds, viii, x, xii, 2, 3, 6, 8, 12, 14, 32, 45, 55, 56, 60, 70, 75, 76, 78, 81, 86, 91, 92, 94, 103, 108, 109, 110, 118, 120, 121, 123, 125, 126

configuration, 46, 48, 51, 53, 55
constituents, 75, 78, 98, 101, 120
consumer, xii, 24, 33, 98, 99
consumers, viii, xii, 2, 4, 22, 24, 38, 98, 99, 110
contamination, 107
controlled trials, 83
coral, v, vii, viii, 1, 4, 5, 6, 16, 21, 24, 25, 26, 28, 29, 30, 33, 35, 36, 37, 38, 39, 41, 55
coral reefs, 5
coronary heart disease, 87
correlation, viii, 2, 25, 35, 37, 80
correlations, 25, 35
corticosteroids, 3
cosmetics, xii, 118
crop, 60, 86, 87, 98, 100
cultivars, vii, ix, x, 59, 60, 63, 64, 67, 68, 73, 75, 76, 80, 82, 84, 99, 108, 125, 132
cycloartenol, 21, 27, 28, 29, 31, 35

D

data set, 63, 68
data structure, 91
decomposition, 9, 121
degradation, 68, 79, 90, 108, 109, 131
dehydrodinosterol, 4
Demospongiae, ix, 43, 44, 56
derivatives, 14, 46, 100, 122, 123, 126
desmethylsterol, 101, 102, 103, 104
desmosterol, 4, 20, 30
detection, xi, 60, 70, 87, 95, 98, 100, 105, 107, 108, 109, 110, 112, 114, 118, 122, 123, 124, 131
detection techniques, 122
Dictyota sp., 26, 27, 28, 29, 34, 35
diet composition, 37
dihydrocholesterol, 22
dimethylsterol, 101, 107
dinoflagellates, 4

dinosterol, 4, 21, 27, 28, 31, 32
discriminant analysis, 63, 70, 95, 105
discrimination, 60, 62, 63, 68, 132, 133
distinctive sterol(s), 24, 28, 29, 34, 38
distribution, 7, 101, 102
diversity, ix, 41, 43, 44, 45, 57, 71, 75, 108, 126
domestication, 75
double bonds, viii, 2, 5, 15, 17, 22, 26, 109
drug discovery, 46, 57

E

E. lucunter, 28, 29, 30, 31, 34, 37
electric current, 8
electric field, 120
electron, 10, 14, 124
elucidation, 125
employment, 123
energy, 2, 9, 10, 47, 57
environment, 44, 45, 76, 120
environmental conditions, 106
environmental factors, 135
environments, ix, 43, 44
episterol, viii, 2, 4, 20, 27, 28, 29, 30, 32, 33, 34, 35, 38
ergosterol, 4, 5, 20, 30, 47, 120, 124
erythrodiol, xi, 98, 102, 103, 108, 111, 115
ethanol, 51, 120, 121, 136
ethers, viii, 1, 3, 12, 14, 15, 18, 122, 123
ethyl acetate, 51
European Union, xii, 65, 77, 88, 98, 99, 113
experimental condition, 123
extraction, viii, xii, 95, 100, 101, 102, 103, 105, 106, 111, 112, 118, 119, 120, 121, 128, 129, 135
extracts, 17, 48, 56, 135

F

fat, 3, 5, 131

fatty acid methyl esters, 6, 13
fatty acids, xi, 2, 3, 6, 24, 39, 45, 46, 63, 68, 86, 93, 98, 107, 118, 119, 121, 126, 134
FID, 2, 6, 7, 8, 17, 19, 20, 21, 38, 65, 77, 88, 109, 123, 126
fish, viii, 2, 12, 22, 24, 29, 32, 33, 35, 36, 37, 38, 40
flame, viii, 2, 6, 7, 8, 65, 77, 88, 123, 131
flame-ionization detectors, 6
fluid extract, 120, 128
food, vii, viii, xii, 1, 2, 4, 5, 25, 26, 37, 38, 39, 40, 44, 86, 98, 110, 117, 118, 119, 130, 134
food products, xii, 98, 118
food web, vii, viii, 1, 2, 4, 25, 26, 37, 38, 39, 40
formation, 8, 10, 105
formula, 26, 27, 30, 31, 32
freshwater, vii, ix, 39, 43, 44, 45, 46, 47, 48, 50, 51, 52, 53, 54, 55, 56, 57
freshwater sponges, 44, 45, 46, 48, 51, 52, 53, 55, 56, 57
fruits, 64, 67, 79, 86, 88, 102, 106, 136
fucostanol, 4

G

Galaxaura sp., 26, 27, 28, 29, 34, 36
gas chromatography, vii, viii, 1, 6, 40, 41, 47, 52, 65, 77, 78, 88, 113, 114, 122, 130
GC, 2, 6, 7, 10, 17, 19, 20, 21, 38, 39, 53, 109, 112, 113, 114, 115, 118, 122, 123, 124, 126, 129, 132, 133
GC-FID, 6, 7, 17, 19, 20, 21, 38, 126
GC-MS, 2, 6, 7, 10, 17, 19, 20, 21, 38, 53, 123, 124, 126
gel, 48, 51, 65, 77, 88, 121
genetic factors, 69, 93
genomics, 40
geographical origin, 70, 94, 105, 125, 132
glycerol, 2

God, 94
gorgosterol, viii, 2, 4, 21, 27, 28, 29, 31, 33, 34, 35, 37, 38
grades, 103
grass, 4
grazers, 24
Greece, 72, 74, 115
green alga, 4, 25, 28, 29, 33, 38
growth, 22, 37, 41, 96
Gulf of Mexico, 24, 39

H

H. opuntia, 25, 26, 27, 28, 29, 34, 35, 36
habitat, 37
halophyte, 135
harvesting, 64, 76, 88
health, vii, ix, x, xii, 41, 59, 78, 85, 89, 90, 98, 99, 117, 118, 119, 120, 126
health effects, 118
helium, 6, 7, 8, 65, 77, 88
heterogeneity, 99
hexane, 6, 18, 51, 53, 102, 120, 121, 124
hogfish, 29, 33, 35, 36
hormones, 5
host, 22, 24
HPLC, 48, 51, 102, 109, 114, 118, 123, 124, 125, 131
human, xii, 4, 5, 78, 86, 87, 95, 96, 115, 117, 119, 124, 126, 128
human health, xii, 78, 86, 87, 117, 119, 124, 126, 128
hydrocarbons, 6, 75, 102, 121
hydrogen, 6, 8, 12, 13, 14, 15
hydrolysis, 2, 121
hydrophobicity, 135
hydroxide, 65, 77, 88, 121
hydroxyl, 3, 100, 118

Index

I

identification, viii, xii, 2, 6, 7, 15, 19, 20, 21, 38, 47, 51, 52, 53, 65, 77, 89, 109, 112, 118, 119, 122, 123, 124, 126
inflammatory disease, x, 85
ingestion, 46
ingredients, 128, 130
insects, 22
integration, 18
interface, 10
invertebrates, 24, 29, 37, 40, 45, 47, 56
ionization, viii, 2, 6, 7, 10, 11, 13, 14, 15, 54, 123, 124, 131
irradiation, 135, 136
irrigation, 61, 87
Isofucosterol, 21, 25, 27, 29, 31, 32
isolation, viii, xii, 45, 52, 109, 118, 119, 122, 126, 129
isomerization, 109, 121
Italy, 74

L

landscape, 128
lathosterol, 6, 17, 20, 22, 32
legislation, xii, 68, 78, 79, 91, 98
lipids, viii, 2, 3, 6, 25, 33, 37, 40, 45, 46, 50, 56, 57, 120, 134, 136
liquid chromatography, 6, 48, 70, 95, 111, 125, 129, 130, 131

M

M. cavernosa, 28, 29, 30, 31, 33, 34, 37, 38
magnetic resonance, 125
majority, vii, ix, 45, 47, 50, 59
management, 86
mangrove, 25, 28, 29, 33, 38
mapping, 111
marine environment, vii, ix, 44
masked goby, 29, 33, 35, 36
mass, vii, viii, 1, 6, 7, 9, 10, 11, 12, 15, 16, 17, 40, 41, 47, 52, 53, 54, 57, 70, 95, 112, 113, 123, 124, 125, 130, 132
mass spectra, viii, 2, 10, 16, 17, 41, 52, 53
mass spectrometry, vii, viii, 1, 9, 40, 52, 54, 70, 95, 112, 113, 124, 125, 130, 132
mass spectrum, 9, 12, 15
mass-selection detectors, 6
matrix, 6, 112, 119, 124
methodology, vii, viii, 1
methyl group, 4, 12, 14, 101
methyl groups, 12, 14, 101
methyl groups of sterol-TMS, 13
methylene chloride, 120
microorganisms, 44, 45
molecular identification, 15
molecular ion, 6, 10, 15, 16
molecular mass, 9
molecular structure, 9, 16, 118
molecular weight, 4, 13, 18
molecules, 3, 6, 9, 10
monomethylsterol, 101
MS, 6, 7, 9, 13, 17, 18, 38, 52, 109, 115, 118, 123, 124, 125, 132, 134
multivariate analysis, 69, 83, 105

N

natural compound, 118, 120
neutral, 122, 134
New olive variety, 74
nitrogen, 6, 7, 8, 114
non-polar, 12, 122
North America, 129
Nuclear Magnetic Resonance, 110, 125, 131
nucleus, 22
nutrition, 115

O

occelasterol, 19, 26, 29, 30, 32, 36
oil, vii, ix, x, xi, 5, 59, 60, 61, 62, 63, 64, 65, 67, 68, 69, 70, 73, 74, 75, 76, 77, 78, 79, 81, 82, 83, 85, 86, 87, 88, 89, 90, 91, 92, 93, 94, 95, 98, 99, 100, 101, 103, 104, 105, 106, 107, 108, 109, 110, 111, 112, 113, 114, 115, 116, 119, 125, 127, 133,135
oil samples, x, 61, 64, 67, 68, 74, 91
olive oil, v, vi, vii, ix, x, xi, xii, 59, 60, 61, 62, 63, 64, 66, 67, 68, 69, 70, 71, 72, 73, 74, 75, 76, 77, 78, 79, 80, 82, 83, 84, 85, 86, 87, 89, 90, 91, 93, 94, 95, 96, 97, 98, 99, 100, 102, 103, 104, 105, 106, 107, 108, 109, 110, 111, 112, 113, 114, 115, 116, 125, 132, 133
olive variety, 60, 63, 68
olive-pomace oil, xi, 60, 98, 103, 113
omega-3, 39, 129
organism, 33, 38, 46
organs, 44
ox, 46, 55

P

P. carnea, 28, 29, 30, 31, 33, 34, 37, 38
palm oil, 119
Parque Nacional Sistema Arrecifal Veracruzano, 23
pathway, 14, 101
pattern recognition, 71
peptides, 45, 46
permeability, 100
personal computers, 17
petroleum, 120, 121
phenolic compounds, 75, 82
phospholipids, 2, 25, 33, 112, 121
phylum, 44
physiology, 3, 84

phytoplankton, viii, 2, 4, 25, 26, 27, 28, 29, 33, 34, 35, 36, 37, 38, 126, 133
phytosterols, vi, viii, x, xii, 5, 21, 22, 28, 37, 41, 60, 64, 79, 80, 85, 91, 97, 100, 101, 102, 106, 112, 117, 118, 119, 120, 121, 122, 123, 124, 125, 126, 127, 128, 129, 130, 131, 132, 133
plant sterols, 118, 126, 132
plants, xii, 21, 117, 119, 120, 123, 125, 128
polar, 2, 3, 25, 102, 111, 122, 124, 132
polymorphisms, 83
population, 9
poriferasterol, 4, 21, 27, 28, 29, 31, 33, 34, 35
Portugal, 115
positive correlation, 25, 37
potassium, 45, 65, 77, 88, 121
predators, viii, 2, 24
prevention, 41, 90
primary producer, viii, 2, 5, 24, 25, 26, 28, 34, 37, 38
primary producers, 34
primary products, 2
principal component analysis, 81, 91
probability, 9, 10, 34
producers, viii, 2, 5, 24, 25, 26, 34, 38, 99
pulp, 63, 67, 71, 80, 81, 96, 100
purification, viii, xii, 118, 119, 121, 124

Q

quantification, viii, xii, 19, 65, 77, 88, 109, 118, 119, 122, 124, 129, 131
quantitative estimation, 131

R

R. mangle, 25, 26, 27, 28, 33, 34, 35, 36
Raman spectra, 112
Raman spectroscopy, 110
ramp, 7

Index

reactions, viii, 1, 9, 13, 14
recommendations, iv
recrystallization, 51
eliability, 62
requirements, xii, 117, 119, 124
researchers, 47, 61, 68
resistance, 82
resolution, 8, 47, 48, 123, 124
response, 8, 18
risks, 86

S

safety, 110, 115
science, 111
scope, 56
sea pen shell, 28, 29
sea urchin, 28, 29, 30, 33, 35, 36, 37, 38
sea urchins, 30, 33, 37, 38
seagrass, 25, 28, 29, 33, 38
secondary metabolism, 45
seed, 60, 62, 63, 87, 91, 99, 103, 109, 110, 118, 119, 121, 127, 129
segregation, 63, 68
sensitivity, 8, 111, 123
septic shock, 86
septum, 7
services, iv
showing, 10, 75, 76
side chain, 5, 22, 46, 56, 57, 100, 118
signal transduction, 100
signalling, 3
signals, 3
silica, 45, 48, 51, 65, 77, 88, 102, 121, 124
silicon, 12
sitosterol, x, xi, 4, 5, 21, 22, 27, 31, 32, 33, 61, 64, 65, 66, 67, 74, 77, 78, 79, 80, 89, 90, 94, 98, 101, 102, 106, 118
skeleton, 100, 101, 107
software, 7, 12, 17, 18, 25, 65, 77, 89
solution, 7, 65, 77, 88

Spain, 62, 71, 74, 84
species, vii, viii, 4, 14, 22, 24, 25, 43, 44, 45, 47, 48, 50, 51, 52, 53, 54, 75, 86, 91, 124
spectroscopic techniques, 110
spectroscopy, 110, 113, 131, 132, 134
spinasterol, viii, xi, 21, 27, 28, 86, 90, 91, 92, 94, 118, 127
sponge, v, vii, ix, 28, 29, 33, 35, 36, 37, 43, 45, 46, 47, 48, 50, 51, 52, 53, 54, 55, 56, 57
sponges, viii, 22, 24, 30, 33, 37, 38, 43, 44, 45, 46, 47, 50, 54, 55, 56, 57, 58
stability, 70, 82, 83, 95, 102, 113, 118, 135, 136
stabilizers, 111
standard deviation, 34, 65, 77, 89
stanols, 3, 5, 83, 118, 127
statistics, 26, 32
stellasterol, 4, 20, 26, 28, 29, 30, 32, 35
steroids, 3, 14, 46, 48, 49, 52, 53
sterol, v, vii, viii, ix, x, xi, xii, 1, 4, 5, 7, 12, 15, 18, 19, 21, 22, 24, 25, 26, 27, 28, 29, 30, 31, 32, 33, 35, 36, 37, 38, 40, 41, 44, 45, 46, 48, 49, 51, 53, 54, 56, 57, 59, 60, 61, 62, 63, 65, 66, 67, 68, 69, 70, 71, 72, 73, 74, 76, 77, 78, 80, 81, 83, 84, 86, 87, 88, 89, 91, 95, 96, 98, 101, 102, 103, 104, 105, 106, 108, 109, 112, 115, 117, 118, 121, 123, 127, 129, 131
sterol esters, 4, 118, 121
Sterol-Argan Oil-olive oil-schottenol-spinasterol, 86
stigma, 114
stigmastadiene, 90
stigmasterol, ix, x, 4, 5, 6, 17, 20, 22, 27, 30, 32, 60, 61, 63, 64, 66, 67, 68, 69, 74, 79, 80, 82, 90, 91, 102, 103, 104, 105, 106, 107, 108, 118, 120, 124, 126, 131
storage, viii, 2, 25, 33, 35, 37, 38, 61, 87, 100, 115, 135
strontium, 45

structural lipids, 25, 33, 37
structure, viii, 2, 3, 4, 5, 16, 82, 100, 106, 111, 125
substitutes, 12
substitution, 40
supplementation, 37
synthesis, 67, 81

T

T. testudinum, 25, 26, 27, 28, 34, 35, 36
target, 99, 133
techniques, 53, 100, 109, 120, 124, 125, 126, 128
technology, 70, 83, 94, 100, 108, 111, 135
temperature, 7, 8, 61, 87, 105, 119, 121, 131
thermal stability, 122
tocopherols, xi, 60, 63, 68, 75, 86, 98, 102, 135
traceability, 98, 99, 100, 133
trans-22-dehydrocholesterol, 4, 26, 28, 29, 30, 35, 38
triglycerides, xi, 60, 98
triterpene dialcohols, 62, 63, 68, 102, 103, 114

U

unimolecular ion, 9
uvaol, xi, 98, 103, 108, 111, 115

V

vacuum, 12
vanadium, 45
variables, 63, 68, 91, 100, 110
varieties, vii, x, 61, 62, 63, 66, 67, 68, 71, 72, 73, 74, 78, 80, 81, 90, 98, 115, 125, 129

vegetable oil, xi, xii, 40, 60, 62, 66, 69, 70, 83, 87, 89, 92, 97, 98, 99, 100, 104, 106, 107, 110, 111, 112, 113, 115, 116, 126, 127, 128
virgin olive oil, vii, ix, x, xi, 59, 60, 62, 64, 66, 68, 69, 70, 71, 73, 77, 78, 79, 80, 83, 84, 87, 94, 95, 96, 97, 99, 100, 104, 107, 111, 112, 113, 114, 115, 125, 132

W

water, 5, 9, 44, 51, 64, 76, 88, 120, 121, 124
web, 4
wood, 100
workers, 50
working conditions, 65, 77, 88

Z

zinc, 45
zirconium, 45
zooplankton, viii, 2, 28, 29, 30, 32, 37, 38
zooxanthellae, viii, 2, 4, 25, 26, 27, 28, 29, 33, 34, 35, 36, 38

β

β-sitosterol, viii, ix, x, xi, 18, 25, 29, 33, 34, 36, 60, 61, 63, 65, 67, 68, 69, 74, 77, 78, 79, 80, 82, 85, 89, 91, 98, 102, 103, 104, 105, 106, 120, 124, 126

Δ

Δ7-stigmastenol, xi, 67, 68, 69, 81, 98, 102, 103, 104, 107, 108

Related Nova Publications

A Comprehensive Guide to Aptamers

Editor: Tom Shuster

Series: Life Sciences Research and Development

Book Description: This collection opens with a focus on recent advancements on the development of nucleic acid aptamers as alternative delivery systems for therapeutic oligonucleotides. Additionally, key examples of targeted delivery of the most common nucleic acid therapeutics, including small interfering RNAs, short hairpin RNAs, microRNAs and antisense oligonucleotides for a number of disorders are discussed.

Softcover ISBN: 978-1-53616-293-6
Retail Price: $95

Innovations in Life Science Research

Editors: Rajeshwar P. Sinha, Shashi Pandey-Rei, and Nandita Ghoshal

Series: Life Sciences Research and Development

Book Description: This book is an attempt to keep abreast with the recent innovations made in various fields of life science research. The book encompasses topics ranging from prokaryotic microscopic organisms to eukaryotic higher plants, distributed in fifteen chapters for the benefit of graduate and postgraduate students as well as young researchers, scientists and professionals.

Hardcover ISBN: 978-1-53615-868-7
Retail Price: $230

To see a complete list of Nova publications, please visit our website at www.novapublishers.com

Related Nova Publications

THE DISSIPATIVE MIND: THE HUMAN BEING AS A TRIADIC DISSIPATIVE STRUCTURE

AUTHORS: Salvatore Chirumbolo, Antonio Vella, and Giovanni Vella

SERIES: Life Sciences Research and Development

BOOK DESCRIPTION: In this book, we attempted to build a new epistemological model of life and the human mind taking into account the intriguing model of Prigogine's dissipative structure.

HARDCOVER ISBN: 978-1-53616-793-1
RETAIL PRICE: $230

To see a complete list of Nova publications, please visit our website at www.novapublishers.com